I0071536

iGlobal
Educational Services
Believe.Inspire.Transform.

Pre-Calculus
Study Guide

To order, contact iGlobal Educational Services

PO Box 94224, Phoenix, AZ 85070
Website: www.iglobaleducation.com
Fax: 512-233-5389

Copyright infringement is a violation of Federal Law.

©2017 by iGlobal Educational Services, Austin, Texas.

ISBN-13: 978-1-944346-68-3

All rights reserved. No part of this publication may be reproduced, translated, stored in retrieval system, or transmitted in any way or by any means (electronic, mechanical, photocopying, recording, or otherwise) without prior written permission from iGlobal Educational Services.

Photocopying of student worksheets by a classroom teacher at a non-profit school who has purchased this publication for his/her own class is permissible. Reproduction of any part of this publication for an entire school or for a school system, by for-profit institutions and tutoring centers, or for commercial sale is strictly prohibited.

Printed in the United States of America.

HOW TO USE THIS STUDY GUIDE

iGlobal Educational Services created this study guide to help you review mathematical concepts that may help you increase your knowledge of Pre-Calculus topics.

This study guide should be used to supplement strong and viable curriculum that encourages differentiation for all diverse learners. They can be used at home, in tutoring sessions, or at school.

TABLE OF CONTENTS

INTRODUCTION TO TRIGONOMETRY 7

DETERMINING TRIGONOMETRIC RATIOS 8

ANGULAR MEASUREMENTS 10

RADIAN MEASURE 11

TRIGONOMETRY 13

TRIGONOMETRY OF SPECIAL ANGLES 13

TRIGONOMETRY OF A UNIT CIRCLE 14

TRIGONOMETRY OF EXTREME ANGLES 16

GRAPHING TRIGONOMETRIC FUNCTIONS 18

TRIGONOMETRIC IDENTITIES & EQUATIONS 22

MULTI-ANGLE TRIGONOMETRY 26

CONCLUSION 28

VECTORS 30

   VECTOR ON x −PLANE 30

OPERATION OF VECTORS 31

   QUANTITY AND DIRECTION OF A VECTOR 31

   UNIT VECTOR 32

   PARALLEL VECTORS 33

PROPERTIES OF FUNCTIONS 35

INJECTIVE, SURJECTIVE AND BIJECTIVE 35

ODD AND EVEN FUNCTIONS 36

INCREASING AND DECREASING FUNCTIONS 36

LOCAL MINIMA AND MAXIMA 38

POLYNOMIALS 41

OPERATION OF POLYNOMIALS 41

   FACTORS OF A POLYNOMIAL 44

   ROOTS OF A POLYNOMIAL 44

ABSOLUTE VALUE FUNCTIONS, RATIONALS, EXPONENTS, LOGARITHMS 46

RATIONAL FUNCTIONS 48

   EXPONENTS 50

ALGEBRA REVIEW, PARENT FUNCTIONS, ALGEBRAIC TRANSFORMATIONS, INVERSE FUNCTIONS 52

    PARENT FUNCTION 52

    ALGEBRAIC TRANSFORMATIONS 52

    INVERSE FUNCTION 53

CONCLUSION **55**

CONICS, CORDINATE SYSTEM, SERIES 56

CONICS **57**

    CIRCLE 58

    PARABOLA 58

    ELLIPSE 59

    HYPERBOLA 60

POLAR AND PARAMETRIC **64**

    POLAR EQUATIONS 66

SEQUENCES, SERIES AND LIMITS **70**

    SERIES 70

CONCLUSION **74**

GLOSSARY **75**

PRE-CALCULUS: TRIGONOMETRY

WHAT YOU NEED TO KNOW

In this session, we will discuss the abstract concepts of trigonometry. Since it deals with the angles, we will first introduce the basic and commonly used angle measures that is the degrees, second, minutes and the radians. We will then look at the basic trigonometric ratios and their relation. We will move ahead and look at trigonometry of the unit square which is an important tool if determining the trigonometric ratios of angles more than 90°, not positive as well as angles at the extremes. This will have prepared us to look at the trigonometric functions with special attention to graphing the functions and solving them using algebraic rules and trigonometric identities.

MATH TOPICS

- Pre-calculus 103.1 Introduction (Reference 1.11).
- Pre-calculus 103.2 Trigonometry (Reference 1.12).
- Pre-calculus 103.3 Graphing Trigonometric Functions (Reference 1.13).
- Pre-calculus 103.4 Trigonometric Identities and equations (Tie the link to Calculus). (Reference 1.14).
- Pre-calculus 103.5 Multi-angle Trigonometry. (Reference 1.15).

INTRODUCTION

The concept of using triangles in solving day to day life is very common. For instance, the use of inclined planes, defining locations of places using bearing and distance, survey among others. Some of these problems cannot be solved without the use of the triangles ratios called trigonometric ratios. Thus, there is need to understand what are they and how they are applied.

INTRODUCTION TO TRIGONOMETRY

Trigonometry is a topic that deals with ratios of sides in a right triangle with reference to a given angle. The most basic trigonometric ratios are the sine (*sin*), the cosine (*cos*) and the tangent (*tan*). Given a right triangle MNL with an angle, θ, define side ML as opposite side and MN as the adjacent side (with respect to angle θ)

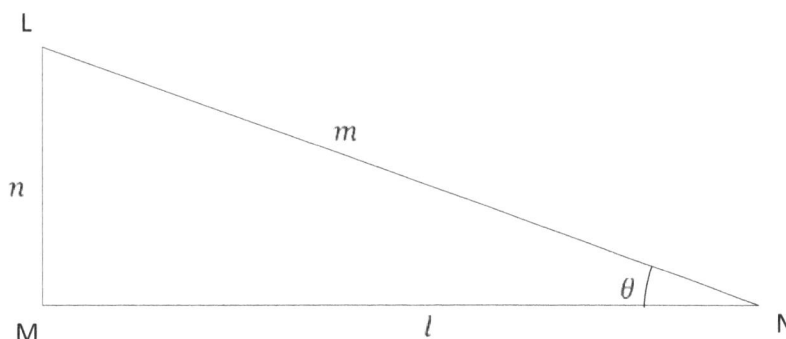

Therefore, we can define the basic ratios as

$$\cos \theta = \frac{l}{m}; \ \sin \theta = \frac{n}{m} \text{ and } \tan \theta = \frac{n}{l}$$

Note that $\qquad \left(\frac{n}{m}\right) \div \left(\frac{l}{m}\right) = \frac{n}{m} \times \frac{m}{l} = \frac{n}{l} = \tan \theta$

Thus. $\qquad \tan \theta = \dfrac{s \quad \theta}{c \quad \theta}$

Taking the reciprocals of the basic trigonometric ratios, we get other ratios; the secant (sec), the cosecant (csc) and the cotangent (cot).

$$\sec\theta = \frac{1}{\cos\theta} = \frac{m}{l}; \ \csc\theta = \frac{1}{\sec\theta} = \frac{m}{n}; \ \cot\theta = \frac{1}{\tan\theta} = \frac{\cos\theta}{\sin\theta} = \frac{m}{l}$$

DETERMINING TRIGONOMETRIC RATIOS

Trigonometric ratios can be determined using the measurements of the appropriate sides given as well as using calculators and trigonometric tables. We will base our discussions on using calculators and the first method only.

The first method is simply using substitution for the appropriate measurement of the hypotenuse, adjacent side or opposite side where applicable.

The second method is the use of the calculator given the degree so as to read out the ratio. The procedure is to press the trigonometric ratio, then enter the value in degrees mode then press equals to.

Example for c₁   30°,
Ensure you are using the degrees mode; press cos button + input 30 + press = button

Example 1
Given that $\sin\alpha = \frac{1}{5}$, find $\cos\alpha$, $\tan\alpha$ and $\sec(90 - \alpha)$.

Solution
By definition, $\sin\alpha = \frac{2}{5} = \frac{O}{H} \quad \frac{S}{S}$

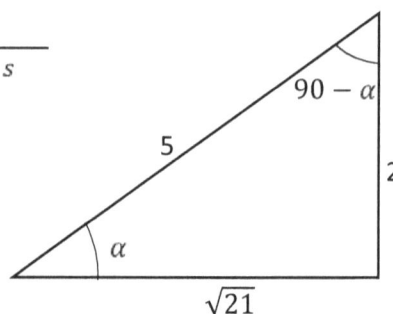

$$90 - \alpha$$
$$5$$
$$2$$
$$\alpha$$
$$\sqrt{21}$$

Adjacent side is $\sqrt{5^2 - 2^2} = \sqrt{21}$

Thus

$$\cos\alpha = \frac{A}{H} \quad\frac{S.}{} = \frac{\sqrt{21}}{5},$$

$$\tan\alpha = \frac{O}{A} \quad\frac{S.}{S.} = \frac{2}{\sqrt{21}} = \frac{2\sqrt{21}}{21}$$

$$\sec(90 - \alpha) = \frac{H}{A} \quad\frac{}{S.} = \frac{5}{2}$$

Example 2

Given that $\tan 30° = 0.5774$, find the hypotenuse of the triangle if the adjacent is 12.5 in.

<u>Solution</u>

By definition,

$$\tan\alpha = \frac{O}{A} \quad\frac{S.}{S.}$$

Upon substitution, we have

$$\tan 30° = \frac{O.}{12.5} \quad\frac{S.}{} = 0.5774$$

Thus, the opposite side is $12.5 \times 0.5774 = 7.218$ in

Using the Pythagorean Theorem, we have the hypotenuse = $\sqrt{7.218^2 + 12.5^2} = 14.43$ in

ANGULAR MEASUREMENTS

An angle can be measured in degrees or radians. When we use degrees as a mean of measurement, we may use represent small angular measurements of degrees in terms of minutes (abbreviated as ') and seconds (abbreviated as "). The system of representing angular measurement in terms of degrees, minutes and second is sometimes referred to as Degree-Minutes-Seconds system (DMS). We use the following conversations

$$1° = 60' \text{ and } 1' = 60''$$

$$\text{Thus, } 1° = 3600''$$

Example 3

Find the size of the angle β in DMS system in the following triangle

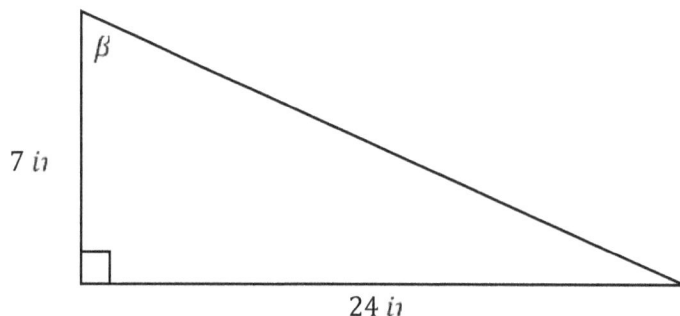

Solution

With respect to angle β, the sides whose measurement are given are

$a \qquad s_i \qquad = 7 \text{ in} \quad \text{and } o \qquad\qquad s_i \qquad = 24 \text{ in}$, hence, we will use tangent.

$$\tan \beta = \frac{24}{7} = 3.429$$

Thus, we find an angle whose tangent is 3.429. This is called the tangent inverse of 3.429 denoted $\tan^{-1} 3.429$

Using calculator, we press shift + tan key + input 3.429 + equals to key
$$\text{Thus, } \beta = \tan^{-1} 3.429 = 73.74°$$

We convert $0.74°$ to minutes and second respectively.

Into minutes, we will have $0.74° = 0.74 \times 60 = 44.4'$

We then convert 0.4' to seconds. Thus, $0.4' = 0.4 \times 60 = 24''$

Thus, $\beta = 73°\,44'\,24''$

RADIAN MEASURE

A radian is equal to a central angle whose arc length is equal to the radius. Since the circumference is given by 2π , there are $\dfrac{2\pi}{r} = 2\pi$ in one cycle.

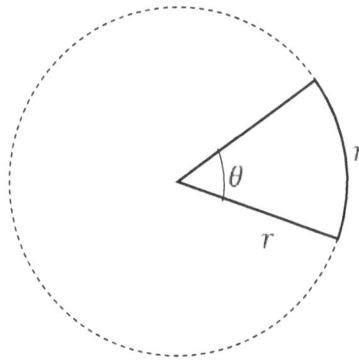

To convert an angle in degrees to radians, we simply multiply it by $\dfrac{\pi}{1}$ while to convert a radian to degree, we multiply it by $\dfrac{1}{\pi}$.

Thus,

$$60° = 60 \times \frac{\pi}{180} = \frac{\pi}{3}\,r$$

$$\frac{4}{3}\pi\,r \qquad = \frac{4}{3}\pi \times \frac{180}{\pi} = 240°$$

Example 4

Convert the following to radians in terms of π.

(i). 2 $^\circ$ (ii). 3 $^\circ$ (iii). 1 $^\circ$

Solution

To convert an angle in degrees to radians we multiply to value by $\frac{\pi}{1}$.

(i). $270\,d$ $= 270 \times \dfrac{\pi}{1} = \dfrac{2}{3}\pi\,r$

(ii). $300\,d$ $= 300 \times \dfrac{\pi}{1} = \dfrac{5}{3}\pi\,r$

(iii). $15\,d$ $= 15 \times \dfrac{\pi}{1} = \dfrac{\pi}{1}\,r$

Example 5

Convert the following to degrees

(i). $\dfrac{3}{4}\pi$ (ii). $\dfrac{\pi}{4}$ (iii). $\dfrac{\pi}{5}$

Solution

To convert an angle in degrees to radians we multiply to value by $\frac{1}{\pi}$.

(i). $\dfrac{3}{4}\pi = \dfrac{3}{4}\pi \times \dfrac{1}{\pi} = 135^\circ$

(ii). $\dfrac{1}{4}\pi = \dfrac{1}{4}\pi \times \dfrac{1}{\pi} = 45^\circ$

(iii). $\dfrac{1}{5}\pi = \dfrac{1}{5}\pi \times \dfrac{1}{\pi} = 36^\circ$

TRIGONOMETRY

In this topic, we are concerned with of special angles, the negative angles and angles more than 90°. This is a concept that will enable us solve quite a number of trigonometric problems whose scope of $\cos 60° = \frac{1}{2}$; $\sin 60° = \frac{\sqrt{3}}{2}$; $\tan 60° = \sqrt{3}$, angle is more than 90°. On the other side, it will also enable us write the specific trigonometric ratio to a given angle.

TRIGONOMETRY OF SPECIAL ANGLES

The special angles here are $30°, 45°$ and $60°$. To achieve the objective of this subtopic, we consider two right triangles, a $30 - 60 - 90$ and a $45 - 45 - 90$ right triangles.

$$30 - 60 - 90$$

Consider half of an equilateral triangle of side 2 units below

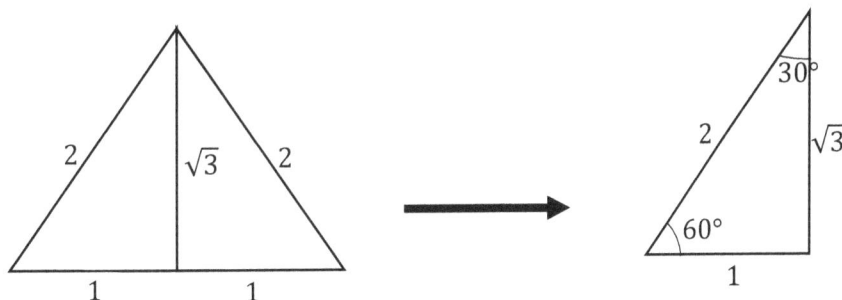

From the triangle, we have

$$\cos 60° = \frac{1}{2}; \ \sin 60° = \frac{\sqrt{3}}{2}; \ \tan 60° = \sqrt{3}$$

$$\cos 30° = \frac{\sqrt{3}}{2}; \ \sin 30° = \frac{1}{2}; \ \tan 30° = \frac{1}{\sqrt{3}}$$

Note that $30° = \frac{\pi}{6}$ and $60° = \frac{\pi}{3}$

TRIGONOMETRY OF A UNIT CIRCLE

Most trigonometric ratios of most angles can be calculated using the concept of trigonometric ratios. In this case, a unit circle (circle of radius 1) with center at the origin is considered. It is divided into four parts called quadrants. Each quadrant has its own procedure which should be used to find the trigonometric ratio of an angle within it. The quadrants are numbered from the positive horizontal axis gaining counter clockwise. Angles read in this direction are said to be positive while those read from the clockwise direction from the positive x −axis are the negative angles.

When we read angles from positive x −axis as 0°, in counter clockwise direction, positive y −axis would be 90°, negative x −axis 180°, negative y −axis 270° and positive x −axis after the turn is 360°. If we continue in that manner, after another turn, positive $x − a$ would be 720°, positive y −axis would be 810° (after two turns) and soon.

For instance, the figure below shows rays at approximately 100° and another at −100°.

Since −100°, is read in anticlockwise direction from positive x −axis, its positive equivalent angle is 360° + (θ), here $\theta = -100°$. Thus, we have 360° + (−100) = 260°

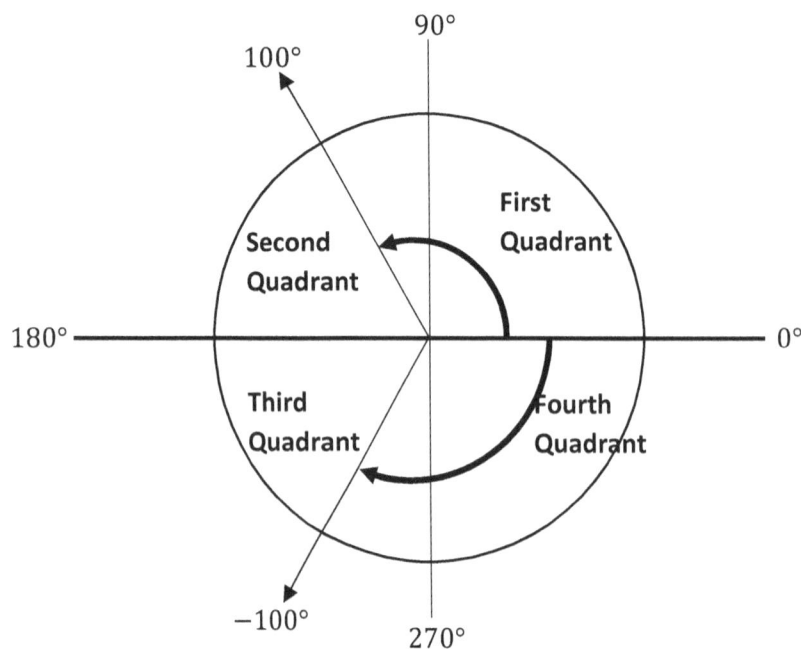

Trigonometric ratios are best read using acute angles, therefore, for angles more than 90°, we get an equivalent of it.

If θ is

(i). In the first quadrant, we read it directly from tables and calculator
(ii) In the second quadrant, its equivalent is $180° - \theta$
(iii) In the third quadrant, its equivalent is $\theta - 180°$
(iv) In the fourth quadrant, its equivalent is $360° - \theta$

In short, we measure the absolute of an angle from the $x - a$, either from the positive axis or negative axis so along as the resultant is acute.

If the angle is more than 360°, we subtract multiples of 360 till it becomes less than 360 then apply the above 4 rules.

If the angle is less than 360°, we add multiples of 360 till it becomes less than 360 then apply the above 4 rules.

Our discussed will be based on the basic trigonometric ratios only.
- In the first quadrant, **All ratios are positive**.
- In the second quadrant, **sine alone is positive**.
- In the third quadrant, **tangent alone is positive**.
- In the fourth quadrant, **cosine alone is positive**.

Going by what is positive, we have the short name **CAST** or **ACTS** as shown

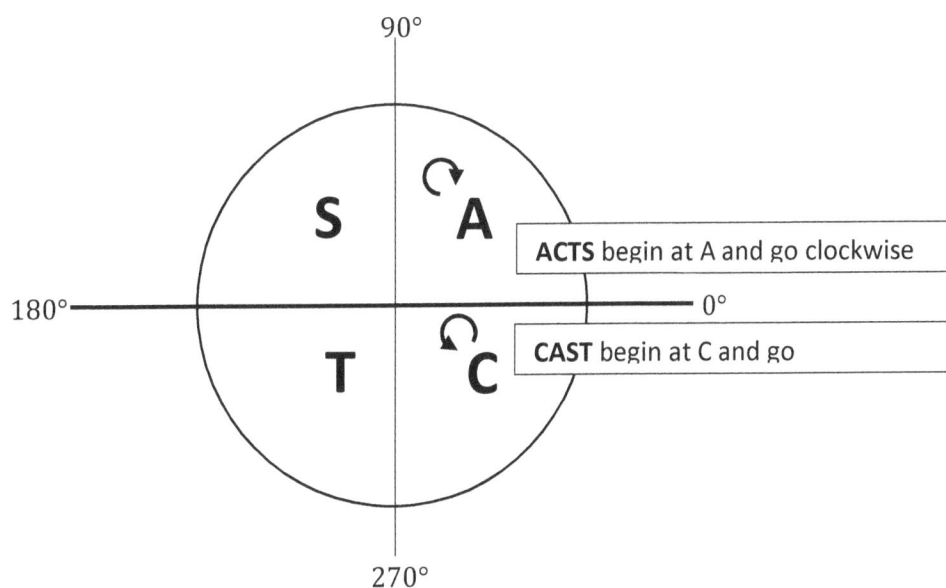

ACTS begin at A and go clockwise

CAST begin at C and go

TRIGONOMETRY OF EXTREME ANGLES

These are angles at the intersection of the quadrants. They are $0°, 90°, 180°, 270°$ and $360°$. We will give a summary of their trigonometric ratios

Angle $0°$ or $360°$
$$\cos 0° = 1, \sin 0° = \tan 0° = 0$$

Angle $90°$
$$\sin 90° = 1, \cos 90° = 0, \ \tan 90° = u \qquad (\infty)$$

Angle $180°$
$$\cos 180° = -1, \sin 0° = \tan 0° = 0$$

Angle $270°$
$$\sin 270° = -1, \cos 270° = 0, \ \tan 270° = u \qquad (\infty)$$

Example 1
Evaluate the following

(i). $c \quad 3 \quad °$ (ii). $t_{\cdot} \quad 3 \quad °$

(iii). $c \quad 9 \quad °$ (iv). $s_{\cdot} \quad 2 \quad °$

<u>Solution</u>

(i). $32°$ is in the first quadrant hence, we read its cosine direct from the table or calculator
$$\cos 32° = 0.8480$$

(ii). $300°$ is in the fourth quadrant hence its equivalent is $360 - 300 = 60°$
Furthermore, $t_{\cdot}$ is negative in that quadrant, hence we have
$$\tan 300° = -\tan 60° = 1.732$$

(iii). $95°$ is in the second quadrant hence its equivalent angle is $180 - 95 = 85°$

Cosine is negative in the second quadrant, hence
$$\cos 95° = -\cos 85 = -0.08716$$

(iv). $220°$ is in the third quadrant, hence its equivalent is $220 - 180 = 40°$

Tangent is positive in the third quadrant hence
$$\tan 220° = \tan 40° = 0.8391$$

Example 2
Evaluate the following

    i. $\sin 30° \tan 45° - \sin 270°$

    ii. $\cos 45° \div \sin 60° \cos 180°$

Solution

(i). From the trigonometry of special angles and extremes, we have

$$\sin 30° = \frac{1}{2}\ ,\tan 45° = 1 \text{ and } \sin 270° = 1$$

Thus, $\sin 30° \tan 45° \times \sin 270° = \frac{1}{2} \times 1 - 1 = \frac{1}{2} - 1 = -\frac{1}{2}$

(ii). From the trigonometry of special angles and extremes, we have

$$\cos 180° = -1\ , \cos 45° = \frac{1}{\sqrt{2}} \text{ and } \sin 60° = \frac{\sqrt{3}}{2}$$

$$\cos 45° \div \sin 60° \cos 180° = \frac{1}{\sqrt{2}} \div \left(\frac{\sqrt{3}}{2} \times -1 \right) = \frac{1}{\sqrt{2}} \times -\frac{2}{\sqrt{3}}$$

$$= -\frac{2}{\sqrt{6}} = -\frac{2\sqrt{6}}{\sqrt{6}\sqrt{6}} = -\frac{2\sqrt{6}}{6} = -\frac{\sqrt{6}}{3}$$

GRAPHING TRIGONOMETRIC FUNCTIONS

Trigonometric functions are algebraic equations where the variable is the angle. For instance, we may say $f(x) = \sin x$ where x changes (variable). To draw a graph of a function, we come up with a set of values for of the input then use them to generate another set of values (the outputs). The ordered pairs for each will help us graph the functions given.

For instance, given the function $f(t) = \sin t \quad a \quad f(t) = \cos t$

We come up with the set of inputs, say

$$-180°, -120°, -90° - 60°, 0°, 60°, 90°, 120°, -180°$$

These are values for
t, we can come up with a table of values having that of $f(t)$ too. Thus

t	−180°	−120°	−90°	−60°	0°	60°	90°	120°	180°
$f(t)$ $= \sin t$	0	- 0.866	-1	- 0.866	0	0.866	1	0.866	0
$f(t)$ $= \cos t$	-1	-0.5	0	0.5	1	0.0.5	0	-0.5	-1

Upon graphing of the common axes, we get the following graph

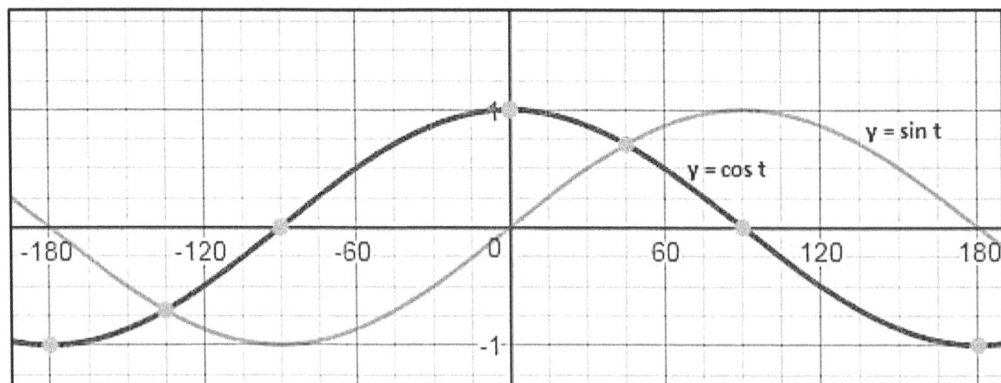

The graph of a general trigonometric function is given by (for cosine) $f(x) = A\cos(w + \theta) + B$ where, A is the amplitude of the graph, w is the angular frequency, θ is the phase angle, $-\dfrac{\theta}{w}$ the phase shift (horizontal shift) and, B is vertical shift and $\dfrac{2\pi}{w}$ the period.

For instance, if we have $f(t) = 2 \sin\left(2t + \frac{\pi}{2}\right) + 1$

Then we can draw the graph of the function above using either of the two methods.

Method 1:
For the set of values of t, come up with the $f(t)$, then graph it. This method was illustrated above.

Method 2:
Having drawn the first one, simply identify the parameters involved then simply shift or expand/contract the original graph $f(t) = \sin t$ to get the required graph.

Comparing the graph with the general function $f(x) = A \sin(w + \theta) + B$, we have

The amplitude (distance from c li to highest point of the graph) , A = 2 units

Angular frequency (number of times the waves makes complete circles in every 2π interval), $w = 2$

Phase shift (Number of units the graph moves horizontally from the parent graph) is

$\left(-\frac{\pi}{2}\right) \div 2 = -\frac{\pi}{4}$ which is a   0.7854 units   to   the   left   vertical shift(distance moved vertically)=1

Period (number of times a cycle is completed) $=\frac{2\pi}{w} = \frac{2\pi}{2} = \pi$

Upon drawing the graph we have

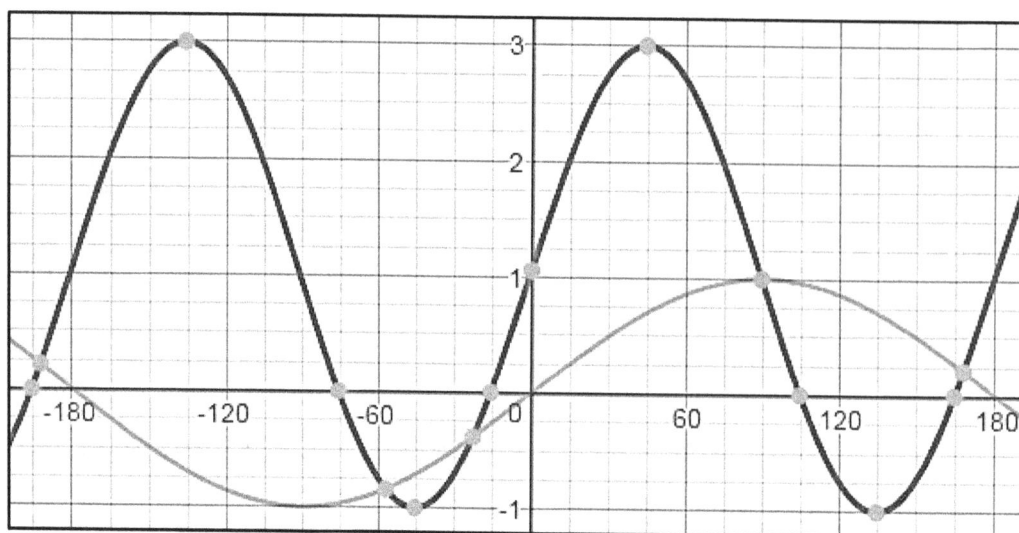

$$y = 2\,si\ \left(2\ +\frac{\pi}{2}\right) + 1$$

$$y = si\ \ t$$

Example 1

Draw the graph of $y = \frac{1}{2}\cos 3t + 2$ for $-\pi < t < \pi$ in step of $\frac{\pi}{6}$.

Solution

We come up with a table of values of y for each t.

t	$-180°$	$-120°$	$-60°$	$0°$	$60°$	$120°$	$180°$
y $= 0.5\cos 3t$ $+ 2$	1.5	2.5	1.5	2.5	1.5	2.5	1.5

Upon graphing, we get

$$y = 0.5\,c \quad (3\ ) + 2$$

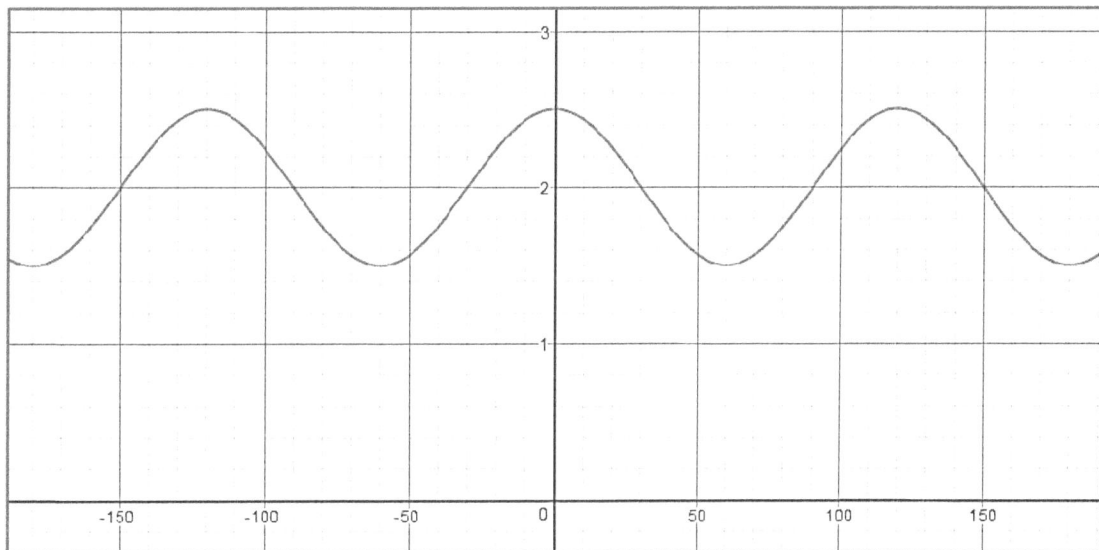

TRIGONOMETRIC IDENTITIES & EQUATIONS

Trigonometric identities are a number of trigonometric rules set to help solve trigonometric expressions and equations. They are composed of basic as well as non-basic trigonometric ratios. In this lesson, we will state the most common ones and use them to solve and simplify expressions and equations.

PYTHAGOREAN IDENTITIES

$$\cos^2 x + \sin^2 x = 1$$

$$1 + \tan^2 x = \sec^2 x$$

$$\cot^2 x + 1 = \csc^2 x$$

SUM AND DIFFERENCE OF ANGLES

$$\sin(x \pm y) = \sin x \cos y \pm \cos x \sin y$$

$$\cos(x \pm y) = \cos x \cos y \mp \sin x \sin y$$

$$\tan(x \pm y) = \frac{\tan x \pm \tan y}{1 \mp \tan x \tan y}$$

DOUBLE ANGLE

$$\cos 2x = \cos^2 x - \sin^2 x$$

$$\sin 2x = 2 \sin x \cos x$$

$$\tan 2x = \frac{2 \tan x}{1 - \tan^2 x}$$

SQUARE IDENTITIES

$$\cos^2 x = \frac{1}{2}(1 + c \quad 2x)$$

$$\sin^2 x = \frac{1}{2}(1 - c \quad 2x)$$

$$\tan^2 x = \frac{1 - \cos 2x}{1 + \cos 2x}$$

PRODUCT TO SUM FORMULAS

$$\sin x \sin y = \frac{1}{2}(\cos(x-y) - \cos(x+y))$$

$$\cos x \cos y = \frac{1}{2}(\cos(x-y) + \cos(x+y))$$

$$\sin x \cos y = \frac{1}{2}(\sin(x+y) + \sin(x-y))$$

$$\cos x \sin y = \frac{1}{2}(\sin(x+y) - \sin(x-y))$$

Example 1

1. **Show that** $\frac{1}{2}$ t: $^2 x + c$ $^2 x + \frac{1}{2}$ c 2 t: $^2 x = 1$

Solution

We first collect like terms

$$\frac{1}{2}\tan^2 x + \cos^2 x + \frac{1}{2}\cos 2x \tan^2 x = \frac{1}{2}\tan^2 x + \frac{1}{2}\cos 2x \tan^2 x + \cos^2 x$$

Upon factorization, we have

$$= \tan^2 x \left(\frac{1}{2} + \frac{1}{2}\cos 2x\right) + \cos^2 x$$

$$B \quad \frac{1}{2} + \frac{1}{2}\cos 2x = \cos^2 x$$

Upon substitution, we have

$$= \tan^2 x \cos^2 x + \cos^2 x$$

$$S \quad \tan^2 x = \frac{\sin^2 x}{\cos^2 x}, \tan^2 x \cos^2 x + \cos^2 x = \frac{\sin^2 x}{\cos^2 x}\cos^2 x = \sin^2 x$$

Thus, $\tan^2 x \cos^2 x + \cos^2 x = \sin^2 x + \cos^2 x = 1$

2. **Show that** $\dfrac{\cos^2 x}{1+\tan^2 x} - \dfrac{1-\cos 2x}{2+2\tan^2 x} = \dfrac{1}{2}\cos 2x + \dfrac{1}{2}\cos^2 2x$

Solution

Since $1 + \tan^2 x = \sec^2 x$, we carry out substitution to get

$$\frac{\cos^2 x}{1 + \tan^2 x} - \frac{1 - \cos 2x}{2 + 2\tan^2 x} = \frac{\cos^2 x}{\sec^2 x} - \frac{1 - \cos 2x}{2\sec^2 x}$$

$$= \frac{\cos^2 x}{\sec^2 x} - \frac{\frac{1}{2}(1 - \cos 2x)}{\sec^2 x}$$

Since $\frac{1}{2}(1 - \cos 2x) = \sin^2 x$, we can substitute into the above expression to get

$$\frac{\cos^2 x}{\sec^2 x} - \frac{\frac{1}{2}(1 - \cos 2x)}{\sec^2 x} = \frac{\cos^2 x}{\sec^2 x} - \frac{\sin^2 x}{\sec^2 x} = \frac{\cos^2 x - \sin^2 x}{\sec^2 x}$$

Since $\cos^2 x - \sin^2 x = \cos 2x$, we make substitution to get

$$\frac{\cos^2 x - \sin^2 x}{\sec^2 x} = \frac{\cos 2x}{\sec^2 x}$$

But $\sec^2 x = \dfrac{1}{\cos^2 x}$

Thus, we expression changes

$$\frac{\cos 2x}{\sec^2 x} = \cos 2x \cos^2 x$$

Using the identity $\frac{1}{2} + \frac{1}{2}\cos 2x = \cos^2 x$, we get

$$\cos 2x \cos^2 x = \cos 2x \left(\frac{1}{2} + \frac{1}{2}\cos 2x\right) = \frac{1}{2}\cos 2x + \frac{1}{2}\cos^2 2x$$

TRIGONOMETRIC EQUATIONS

These are algebraic setups where a trigonometric expressions are equated. In this section, we will deal with solutions of trigonometric equations only. This requires the use of trigonometric identities as well as algebraic rules to be able to compute the solutions.

Solution

Since the tank is open, there are only five faces to plaster, that is the side faces and the bottom face.
$$l = 6 f \quad , w = 5 f \quad , h = 4 f$$

The total surface area will be given by

$$\text{Area} = lw + 2wh + 2lh = (6 \times 5) + 2(5 \times 4) + 2(6 \times 4)$$
$$= 30 + 40 + 48$$
$$= 118 \, sq \, .f$$

Example 3

Find the solution of the following equations

(i). $\cos^2 x = -\cos x$

(ii). $\sin x + \dfrac{\cos^2 x}{\sin x} = 0$

Solution

(i). We carry out factorization, $\cos^2 x = -\cos x$ is equivalent to $\cos^2 x + \cos x = 0$

$$\cos^2 x + \cos x = \cos x \, (\cos x + 1) = 0$$

Thus $\cos x = 0$ or $\cos x + 1 = 0$

This implies

$\cos x = 0$ or $\cos x = -1$

For $\cos x = 0, x = 90° = \dfrac{\pi}{2}$.

For $\cos x = -1, x = 180° = \pi$.

(ii). Multiplying by $\sin x$ does not yield desired results, hence, we simplify $\dfrac{c^2 x}{s\ x}$

$$\frac{\cos^2 x}{\sin x} = \frac{\cos^2 x \sin x}{\sin x \sin x} = \frac{\cos^2 x}{\sin^2 x}\sin x = \cot^2 x \sin x$$

Thus,

$$\sin x + \frac{\cos^2 x}{\sin x} = \sin x + \cot^2 x \sin x = \sin x\,(1 + \cot^2 x) = 0$$

Thus, $\sin x = 0$ or $1 + \cot^2 x = 0$

Since $1 + \cot^2 x = \csc^2 x$, we have

$1 + \cot^2 x = \csc^2 x = 0$ implying that $\csc x = 0$

When $\sin x = 0, x = 0°$

When $\csc x = \dfrac{hy_i}{o_1}\ s = 0 = \dfrac{0}{a\ n_1}$

This is undefined since we cannot have such situation.

Hence, the answer is zero.

MULTI-ANGLE TRIGONOMETRY

Multi-angle trigonometry is a concept where a trigonometric problem has a solution that are very many angles. This arises where the variable of the trigonometric function is a function of another variable. T

For instance, when we say, $\cos(2t) = -1$, then the variable of the trigonometric function is $\theta = 2t$. See that θ is a function of another variable t.

To solve such problems, we first determine the solution of the argument of the trigonometric function in all possible quadrants then determine the solution of the final variable, t. Since most trigonometric functions are periodic, there are infinitely many solutions to such problems

Sometimes, we may restrict the solution so that we have one solution or two only.

We will approach to this by examples.

Example 1

Find all the solutions of the following equations
(i). $\cos(2t) = -1$
(ii). $\sin\frac{1}{2}t = 0.5$
(iii). $\cos\frac{2}{3}x = -\frac{\sqrt{3}}{2}$ for $0 < x < 2\pi$

Solution

(i). $\cos(2t) = -1$

Let be $\theta = 2t$, then we have $\cos\theta = -1$, and $\theta = \cos^{-1}-1 = 180°$
In one cycle, $\theta = 180°$
After, 2, 3, 4,... cycles, the angle would be
$\theta = 180°, 540°, 900°, 1260, ...$

We substitute back, to get t; $\theta = 2t, t = \frac{\theta}{2}$
$$t = \frac{180°}{2}, \frac{540°}{2}, \frac{900°}{2}, \frac{1260°}{2}, ...$$

$t = 90°, 270°, 450°, 630°, ...$

(ii). $\sin\frac{1}{2}t = 0.5$

Let be $\theta = \frac{t}{2}$, thus $\sin\theta = 0.5$ and $\theta = \sin^{-1}0.5 = 30°$ in the first quadrant

Since $\sin\theta = 0.5$, is positive, the angle can also be in the second quadrant, hence $\theta = \sin^{-1}0.5 = 30°, 180° - 30° = 150°$

Thus, θ is $30°$ a $150°$ in the first cycle
After the second, third, fourth cycles, we add $360°$ to each angle to get

$$\theta = 30°, 150°, 390°, 510°, 750°, 870°, 1110°, 1230°$$

Since $\theta = \dfrac{t}{2}$;

$$t = 2\theta = 60°, 300°, 780°, 1020°, 1500°, 1740°, 2220°, 2460°$$

(iii). $\cos\dfrac{2}{3}x = -\dfrac{\sqrt{3}}{2}$ for $0 < x < 2\pi$

Let be $\theta = \dfrac{2}{3}x$, thus $\cos\theta = -\dfrac{\sqrt{3}}{2}$ and $\theta = \cos^{-1}\dfrac{\sqrt{3}}{2} = 30°$ in the first quadrant

Since the cosine is negative, the angle is not in the first quadrant but in the second and third quadrant. Thus, $\theta = 180° - 30° = 150°, 180° + 30° = 210°$

Including the equivalent angles in the second cycle, we will have
$\theta = 150°, 210°, 510°, 570°$

Since $\theta = \dfrac{2}{3}x, x = \dfrac{3}{2}\theta = 225°, 315°, 765°, 855°, \ldots$

Since $0 < x < 2\pi$, we have $225° = \dfrac{5}{4}\pi$ a $315° = \dfrac{7}{4}\pi$ as the required answers.

CONCLUSION

In this lesson, we have looked at the abstract concept of trigonometry. We began with the introduction which are concepts of the basic trigonometric ratios and angle measures. We have moved forward to look at various trigonometric ratios of angles that are not acute, positive as well as those that appear special dues to the nature of their representation. We then looked at how we can draw the functions of these graphs and solve them using algebraic and trigonometric identities. The lesson, thus prepared us to look at the application of trigonometry.

VECTORS AND FUNCTIONS

WHAT YOU NEED TO KNOW

In this session, we will discuss basically algebra. This will include a general discussion of functions and other relations followed by a narrow consideration of individual functions. We will also try to see what happens and how to go about manipulation of these functions inform of transformations, determining the inverses among others. It is also important that we consider vectors since it is the basis of describing functions and determining solutions of problems, though, this becomes evident at a letter stage of learning.

MATH TOPICS

- Pre-calculus 103.6 Vectors (Reference 1.11).
- Pre-calculus 103.7 Properties of functions (Reference 1.12).
- Pre-calculus 103.8 Polynomials. (Reference 1.13).
- Pre-calculus 103.9Absolute value functions, Rationals, Exponents, Logarithms (Reference 1.14).
- Pre-calculus 103.10 Algebra review; Parent functions, algebraic transformations, inverse functions (Reference 1.15).

INTRODUCTION

Our daily lives are full of situations where one has to come up with a budget, plan how to spend funds, measure and design site plans among others. In all these activities, one must have some bit of algebra to enhanced problem solving and decision making. Some cases may require that you come up with a formula to solve a problem, but how is this possible if one does not understand algebra. Thus it is vital that we familiarize ourselves with the following topic for easy problem solving processes.

VECTORS

A vector is a geometric entity that has direction as well as quantity. To define direction, we must have the beginning ns the end. Thus, the beginning of a vector is called the tail and the end the head. A vector can be denoted by a single letter with a tilde, a small bold letter, two capital letters with an arrow at their top among others.

The figure below shows vector c, or vector AB written as $\overrightarrow{A}$.

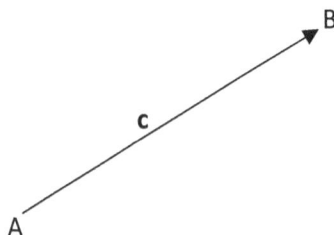

VECTOR ON x —PLANE

A vector on x —plane is represented by $c = \langle s, t \rangle$ or $\binom{s}{t}$ where s and t are called the x — and y — component reactively. If the tail is $A(s_1, t_1)$ and the head is $B(s_2, t_2)$ then the components are x —component, $s = s_2 - s_1$ and the y —component is $t = t_2 - t_1$

Thus, the vector $\overrightarrow{A} = \binom{s_2 - s_1}{t_2 - t_1}$

If the tail of the vector is at the origin, the vector is termed as the **position vector**, while if it is away from the origin, it is called a **column vector.**

OPERATION OF VECTORS

When dealing with vectors, only two operations are defined, other are as a result of manipulation of the two. These are addition and scalar multiplication.

Vector addition
This is done component wise. If we have two vector $a = \langle s_1, t_1 \rangle$ and another vector $b = \langle s_2, t_2 \rangle$, then the sum of these vectors is $a + b = \langle s_1 + s_2, t_1 + t_2 \rangle$

Scalar Multiplication
A scalar is a quantity that has no direction. For instance, a number is a scalar. If k is a scalar and $a = \langle s_1, t_1 \rangle$ is a vector, then scalar multiplication which increases or reduces the side of the vector is given by $k = k\langle s_1, t_1 \rangle = \langle ks_1, kt_1 \rangle$.

QUANTITY AND DIRECTION OF A VECTOR

We need to define these two aspects of vectors so that they can be determined. If we have a vector $c = \langle s, t \rangle$ then the quantity of the vector which implies the length of the vector is given by the Pythagorean Theorem, that is

$$\frac{\|c\| = \sqrt{s^2 + t^2}}{F}$$

The length of a vector is also called the **modulus of a vector.**

If the tail is $A(s_1, t_1)$ and the head is $B(s_2, t_2)$, then the modulus of the vector $\overrightarrow{A}$ is given by

$$\|\overrightarrow{A}\| = \sqrt{(s_2 - s_1)^2 + (t_2 - t_1)^2}$$

The direction of the vector is defined by the angle at which the vector is inclined with respect to the positive x −axis. Thus, the direction of the vector, $c = \langle s, t \rangle$ is given by

$$\theta = t_{\cdot}^{\ -1}\left(\frac{t}{s}\right)$$

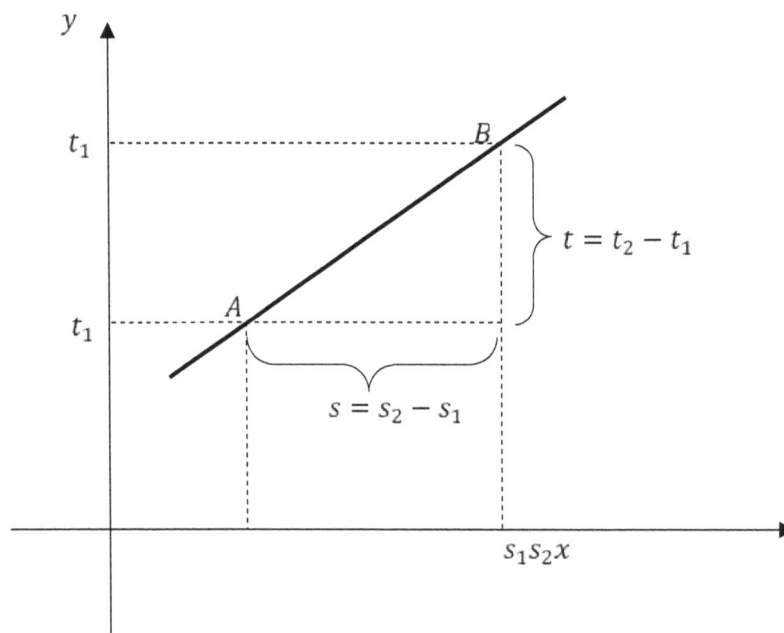

UNIT VECTOR

It is a vector whose magnitude is 1.

Given the vector $c = \langle s, t \rangle$, whose magnitude is $\|c\| = \sqrt{s^2 + t^2}$, the unit vector in its direction is given by

$$\frac{c}{\|c\|} = \sqrt{s^2 + t^2}\langle s, t \rangle = \langle \frac{s}{\sqrt{s^2 + t^2}}, \frac{t}{\sqrt{s^2 + t^2}} \rangle = \begin{pmatrix} \dfrac{s}{\sqrt{s^2 + t^2}} \\ \dfrac{t}{\sqrt{s^2 + t^2}} \end{pmatrix}$$

PARALLEL VECTORS

When vectors are parallel, their numeric representation is a scalar multiple of the other. If we have two vectors $c = \langle s, t \rangle$ and $d = \langle r, m \rangle$ that are parallel, then we must have a scalar, k such that

$$c = \text{kdor } \langle s, t \rangle = k\langle r, m \rangle.$$

Example 1

Find the vector $\overrightarrow{CF}$ given the coordinates $C(0,6)$ and $F(-5,4)$, determine its magnitude.

Solution

$$\overrightarrow{CF} = \begin{pmatrix} -5 - 0 \\ 4 - 6 \end{pmatrix} = \begin{pmatrix} -5 \\ -2 \end{pmatrix},$$

The magnitude would be $\left\| \overrightarrow{CF} \right\| = \sqrt{-5^2 + (-2)^2} = \sqrt{25 + 4} = \sqrt{29}$ sq. units

Example 2

Given that $n = \begin{pmatrix} -1 \\ 2 \end{pmatrix}, m = \begin{pmatrix} 3 \\ 3 \end{pmatrix}$ a $\quad w = \begin{pmatrix} 4 \\ 2 \end{pmatrix}$, Find

(i). $n + m - 3$
(ii). $-2n + 3 \quad + c$
(iii). Find the magnitude and direction of vector $m - 2$

Solution

(i). $n + m - 3c$
$$n + m - 3 = \begin{pmatrix} -1 \\ 2 \end{pmatrix} + \begin{pmatrix} 3 \\ 3 \end{pmatrix} - 3\begin{pmatrix} 4 \\ 2 \end{pmatrix} = \begin{pmatrix} -1 \\ 2 \end{pmatrix} + \begin{pmatrix} 3 \\ 3 \end{pmatrix} - \begin{pmatrix} 12 \\ 6 \end{pmatrix} = \begin{pmatrix} -10 \\ -1 \end{pmatrix}$$

(ii). $-2n + 3m + c$

$$-2n + 3m + c = -2\begin{pmatrix} -1 \\ 2 \end{pmatrix} + 3\begin{pmatrix} 3 \\ 3 \end{pmatrix} + \begin{pmatrix} 4 \\ 2 \end{pmatrix} = \begin{pmatrix} -2 \\ 4 \end{pmatrix} + \begin{pmatrix} 9 \\ 9 \end{pmatrix} + \begin{pmatrix} 4 \\ 2 \end{pmatrix} = \begin{pmatrix} 11 \\ 15 \end{pmatrix}$$

(iii). $m - 2$

$$m - 2 = \begin{pmatrix} 3 \\ 3 \end{pmatrix} - 2\begin{pmatrix} -1 \\ 2 \end{pmatrix} = \begin{pmatrix} 6 \\ -1 \end{pmatrix}$$

The magnitude of the vector is $\|m - 2\| = \sqrt{6^2 + (-1)^2} = \sqrt{36 + 1} = \sqrt{37}$ sq. unit

The direction of the vector is $\theta = t_i{}^{-1}\left(-\frac{1}{6}\right) = t_i{}^{-1} -0.1 = 9.4$

But the angle is not in the first quadrant but in fourth quadrant, hence
$$\theta = 3 \quad - 9.4 \quad = 3 \quad .5$$

Example 3

Find out if the following pairs of vectors are parallel.

(i). $a = \begin{pmatrix} 4 \\ 5 \end{pmatrix}, a \quad b = \begin{pmatrix} 1 \\ 1.5 \end{pmatrix}$

(ii). $c = \begin{pmatrix} -3 \\ -2 \end{pmatrix}, a \quad d = \begin{pmatrix} 1.5 \\ 9.5 \end{pmatrix}$

<u>Solution</u>

(i). If a and b are parallel, then $a = kb$, we find k, if it exists, then vectors are parallel, otherwise, they are not.

$$\begin{pmatrix} 4 \\ 5 \end{pmatrix} = k\begin{pmatrix} 1 \\ 1.5 \end{pmatrix} = \begin{pmatrix} 1 \ k \\ 1.5k \end{pmatrix}$$

Thus, $4 = 14k; k = \frac{4}{1} = \frac{2}{7}$

Also, $5 = 17.5k; \ k = \frac{5}{1.5} = \frac{2}{7}$

Since we get the same answers, $k = \frac{2}{7}$ and it exists, hence the two vectors are parallel.

(ii). If c and d are parallel, then $c = kd$, we find k, if it exists, then vectors are parallel, otherwise, they are not.

$$\begin{pmatrix} -3 \\ -2 \end{pmatrix} = k\begin{pmatrix} 1.5 \\ 9.5 \end{pmatrix} = \begin{pmatrix} 1.5k \\ 9.5k \end{pmatrix}$$

Thus, $-3 = 13.5k; k = \dfrac{-3}{1.5} = 0.2222$

Also, $-2 = 9.5k;\ \ k = \dfrac{-2}{9.5} = 0.2105$

Since we get different answers, k does not exists, hence the two vectors are not parallel.

PROPERTIES OF FUNCTIONS

A function is a relation where every element of the input corresponds to only one members of the output. The collection of all outputs of a function are called the **domain** while that of the output are called the **range**. This implies that we can have two members of the output corresponding to one member of the output, but the vice-versa is not true. To understand function, we will look at the following properties: (i). Injective, surjective and bijective; (ii). Odd and even functions; (iii). Increasing and decreasing functions; and (iv). Local minima and maxima.

INJECTIVE, SURJECTIVE AND BIJECTIVE

A function is injective if there is one to one relation between the elements of the domain and that of the range. That is if $f(a) = f(b)$, then we must have $a = b$. Example of such function is a linear equation.

A function is surjective if for every output, there is a corresponding input. In this situation, the output may be mapped to more than one input. Example $f(x) = x^2$ is subjective when the domain and range are positive real numbers because if $f(x) = 2 = x^2$, we get two element in the output as $\pm\sqrt{2}$. But when the range is the real numbers, then $-2 = x^2$ does not have a real root hence no input for output, -2.

A function is said to be bijectiveif it is both injective and surjective.

ODD AND EVEN FUNCTIONS

A function $f(x)$ is even if substituting x with $-x$ gives $f(x)$. That is $f(-x) = f(x)$. An example of such function is $f(x) = x^2, f(x) = x^4, f = |x|$ and so on. Such functions are symmetric about the y –axis.

A function $f(x)$ is even if substituting x with $-x$ gives $-f(x)$. That is $f(-x) = -f(x)$. An example of such function is $f(x) = x, f(x) = x^3$. Such functions are symmetrical about the origin.

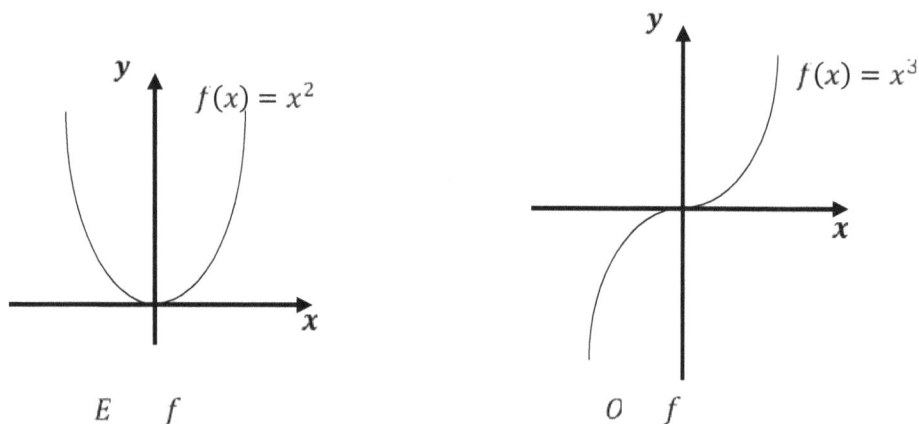

E   f

0   f

INCREASING AND DECREASING FUNCTIONS

A function, $f(x)$ is increasing when for any two values in the domain, a and b such that b is more than a, then the output with respect to b, $f(b)$ must be more than that of $a, f(a)$.

That is, a, b are in the domain of $f(x)$, then $f(b) > f(a)$ when $b > a$. Example $f(x) = 3x + 9$ is increasing. Take $a = 2$ and $b = 5$, clearly $2 = a < b = 5$, then

$$f(2) = 3(2) + 9 = 15; \quad f(5) = 3(5) + 9 = 24$$

Thus, we have $f(5) > f(2)$

A function, $f(x)$ is decreasing if for any two values, c and d such that $c < d$, then the output with respect to c must be more than the output with respect to d. That is, if $c < d$, the $f(c) > f(d)$.

Example, $f(x) = -3x + 9$

Take $c = 1$ and $d = 4$, then $f(1) = -3(1) + 9 = 6$ and $f(4) = -3(4) + 9 = -3$

We found that $f(1) = 6$ is more than $f(4) = -3$.

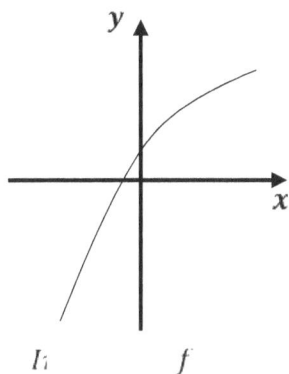

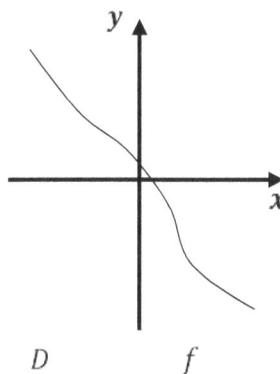

LOCAL MINIMA AND MAXIMA

A function $f(x)$ has a local minima at a point say $x = a$, if $f(x)$ is decreasing for points close and on the left of $x = a$ and increasing for points close and on the right of $x = a$. We can also describe the a local minima using slope. A function has a local minima at $x = a$, if the tangent at the point is horizontal and the value of the functions at points close to a are more than $f(a)$.

A function $f(x)$ has a local maxima at a point say $x = a$, if $f(x)$ is increasing for points close and on the left of $x = a$ and decreasing for points close and on the right of $x = a$. We can also describe the a local minima using slope. A function has a local maxima at $x = a$, if the tangent at the point is horizontal and the value of the functions at points close to a are less than $f(a)$.

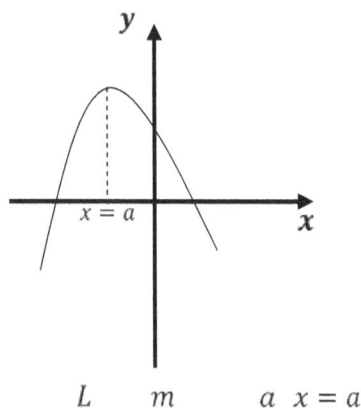

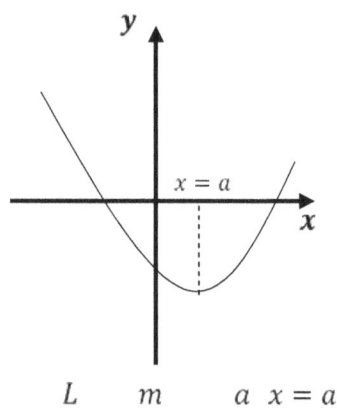

L   m   a $x = a$ L   m   a $x = a$

Example 1

1. Identify the following functions as subjective, injective or objective given that they are defined on the real number system.

   (i). $f(x) = x^3 + 1$

   (ii). $f(x) = x^2 + 2$

   (iii). $f(x) = \dfrac{2}{x+1}$

Solution

(i). Let us have $f(a) = f(b)$, then $f(a) = a^3 + 1 = b^2 + 1 = f(b)$

$a^3 + 1 = b^3 + 1$ Implies $a^3 = b^3$

Taking cube root on both sides, we have $a = b$. Thus, the function is injective.

Let c is a real number in the range of $f(x)$, then we have $x^3 + 1 = c$

or

$x^3 = c - 1$ or

Thus, $x = \sqrt[3]{c - 1}$. But the number $\sqrt[3]{c - 1}$ is a real number since when $c - 1$ is negative, we still get the answer. Thus, the function is surjective.

Therefore, the function is also a bijection since it is injective and surjective.

(ii). Let us have $f(a) = f(b)$, then $f(a) = a^2 + 2 = b^2 + 2 = f(b)$

$a^2 + 2 = b^2 + 2$ Implies $a^2 = b^2$. Taking square roots on both sides, we get $a = \pm\sqrt{b^2}, a = \pm b$

Thus, we have $a = b$ and $a = -b$, hence the function is not injective.

Let c is a real number in the range of $f(x)$, then we have $x^2 + 2 = c$

or

$x^2 = c - 2$ or

Thus, $x = \sqrt[2]{c - 2}$. But the number $\sqrt[2]{c - 1}$ is a not real number since when $c - 1$ is negative, the root is a complex number. Thus, the function is not surjective.

Therefore, the function is neither a bijection.

(iii). Let us have $f(a) = f(b)$, then $f(a) = \frac{2}{a+1} = \frac{2}{b+1} = f(b)$

$2(b + 1) = 2(a + 1)$ or $b + 1 = a + 1$. Hence $a = b$.

The function is injective.

Let c is a real number in the range of $f(x)$, then we have $\frac{2}{x+1} = c$

Then $\frac{x+1}{2} = \frac{1}{c}$ and $x + 1 = \frac{2}{c}$

Thus, $x = \frac{2}{c} - 1$.

If c is zero, x is not defined, hence the function is not surjective, consequently, not bijective.

2. Identify the functions that are odd, even or neither of the two in the following list

(i). $f(x) = x^2 + 3 \ \ + 1$

(ii). $f(x) = -4x^2 + 5$

(iii).

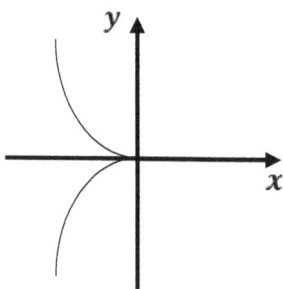

(iv).

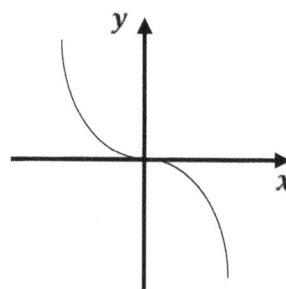

Solution

(i). $f(x) = x^2 + 3x + 1$, substituting $-x$ for x, we get

$$f(-x) = (-x)^2 + 3(-x) + 1 = x^2 - 3x + 1 \neq f(x)$$

Thus, the function is neither odd nor even.

(ii). $f(x) = -4x^2 + 5$, substituting $-x$ for x, we get
$$f(-x) = -4(-x)^2 + 5 = -4x^2 + 5 = f(x)$$

Thus, the function is even

(iii). The graph of the function is neither symmetric about $y - a$ nor the origin, hence it is neither odd nor even.

(iv). The graph of the function is symmetric about the origin, hence it is neither odd.

POLYNOMIALS

A polynomial is a function that is generally given by $p(x) = a_0 + a_1x + a_2x^2 + \cdots + a_nx^n$ where, $a_0, a_1, \ldots a_n$ are called coefficients, $a_n \neq 0$, is called the leading coefficient and x is the variable. The number n which is the highest power in the expression is called the **degree of a polynomial**. The individual parts of the expression separated by the operation, +, are called **terms** and are composed on **a number** and **a variable**. When polynomial is composed of one term, we call it **amonomial,** When it composed of two terms, we call it **a binomial** and so on.

When the degree of a polynomial is 1, we refer to it as a **linear function**, when it is 2, it is a **quadratic function**, when it is 3, we call it **a cubic function** and when it is 4, we call it **a quarticfunction** and so on.

OPERATION OF POLYNOMIALS

Addition and subtraction

Polynomial are added based on like term concept.

If $p_1(x) = a_0 + a_1x + a_2x^2 + \cdots + a_nx^n$ and $p_2(x) = b_0 + b_1x + b_2x^2 + \cdots + b_nx^n$ are two polynomials, then their sum is

$$p_1(x) + p_2(x) = a_0 + b_0 + (a_1 + b_1)x + (a_2 + b_2)x^2 + \cdots + (a_n + b_n)x^n$$

And their difference is

$$p_1(x) - p_2(x) = a_0 - b_0 + (a_1 - b_1)x + (a_2 - b_2)x^2 + \cdots + (a_n - b_n)x^n$$

On addition or subtraction, the degree of the polynomials may decrease however, it cannot be more than that of the highest degree.

Multiplication

Multiplication of polynomials follows the simplification of the product of two expression in brackets. If n and m are the degrees of the two polynomials, then the degree of the resultant polynomial would be n .

Example 1

Find the product of $f(x) = 2x^2 + 3$ and $h(x) = x^3 - 4 + 1$

Solution

The product, $f(x)h(x) = (2x^2 + 3)(x^3 - 4x + 1) = 2x^2(x^3 - 4x + 1) + 3(x^3 - 4x + 1)$

$$= 2x^5 - 8x^3 + 2x^2 + 3x^3 - 12x + 3$$
$$= 2x^5 - 5x^3 + 2x^2 - 12x + 3$$

Division

We use long division to carry divide two polynomials. If the degree of the dividend is n and that of the divisor is m, then the quotient will have a degree of $n - m$ and the remainder will have a degree of less than m.

Example 2

Divide $x^3 + 4x^2 + 7 + 6$ by $x + 2$

<u>Solution</u>

We use long division

At each step of division, we ensure that the highest power disappears when moving to the next step.

We divide x^3 b x to get x^2. Write x^2 above the division sign then multiply the answer by the divisor write the result below x^3 and subtract. The procedure is repeated until we get an expression whose degree is less than that of the divisor.

$$
\begin{array}{r}
x^2 + 2x + 3 \\
x + 2 \overline{\smash{\big)}\ x^3 + 4x^2 + 7x + 6} \\
\underline{-x^3 + 2x^2} \\
2x^2 + 7x \\
\underline{-\ 2x^2 + 4x} \\
3x + 6 \\
\underline{-\ 3x + 6} \\
0
\end{array}
$$

Hence, the quotient is $x^2 + 2x + 3$

Example 2

Divide $2x^5 + 2x^3 + x^2 - 10x + 2$ by $x^2 + 3$

<u>Solution</u>

We use long division

At each step of division, we ensure that the highest power disappears when moving to the next step.

Using the procedure above, we have

$$
\begin{array}{r}
2x^3 - 4x + 1 \\
x^2 + 3 \overline{\smash{\big)} 2x^5 + 0x^4 + 2x^3 + x^2 - 10x + 2} \\
\underline{-2x^5 + 6x^3} \\
-4x^3 + x^2 \\
\underline{- - 4x^3 - 12x} \\
x^2 + 2x \\
\underline{- x^2 + 3} \\
2x - 3 + 2 = 2x - 1
\end{array}
$$

The quotient is $2x^3 - 4x + 1$ while the remainder is $2x - 1$

Hence, we can also write,

$$2x^5 + 2x^3 + x^2 - 10x + 2 = 2x^3 - 4x + 1 + \frac{2x - 1}{x^2 + 3}$$

Multiplying through by $x^2 + 3$, we get

$$
\begin{aligned}
&(2x^5 + 2x^3 + x^2 - 10x + 2)(x^2 + 3) \\
&= (2x^3 - 4x + 1)(x^2 + 3) + (2x + 1)
\end{aligned}
$$

The representation is said to be in division algorithm form.

FACTORS OF A POLYNOMIAL

It is an expression that divides the polynomial completely. If a divisor divided a given polynomial completely, then the quotient together with the divisor are factors of the polynomial, divided.

ROOTS OF A POLYNOMIAL

These are values of a variable which when substituted into the polynomial, it reduces it to zero.

Thus, if $x = t$ is the root of the polynomial, $f(x)$, then $f(t) = 0$.

If $f(x)$ is a polynomial with the root $x = t$, then $x - t$ is factor of the polynomial.

Therefore, If $x = a, x = b$ are allroots of the polynomial, then the polynomial itself would be

$$(x - a)(x - b) = x^2 - (a + b)x + a$$

If the polynomial is quadratic, $y = a + b + c$, the its roots are given by the quadratic formula

$$x = \frac{-b \pm \sqrt{b^2 - 4a}}{2a}$$

Example 3

Find the value of h, if $f(h) = 0$ where $f(x) = 2x^2 + 4x - 1$

Solution

This is a quadratic equation, hence, we use the quadratic formula.

We have $a = 2, b = 4$ and $c = -1$

$$x = \frac{-b \pm \sqrt{b^2 - 4a}}{2a} = x = \frac{-4 \pm \sqrt{16 + 8}}{4} = \frac{-4 \pm 4.899}{4}$$
$$= -1 \pm 1.225$$

$$x = -2.225, \qquad x = 0.225$$

Thus, $h = -2.225$ or $h = 0.225$

ABSOLUTE VALUE FUNCTIONS, RATIONALS, EXPONENTS, LOGARITHMS

Absolute function

This is a function what has all or some of its terms enclosed in absolute symbols, | |. For instance $f(x) = |x|$, $f(x) = |2x - 1| + 4$ and so on. The basic absolute function is $f(x) = |x|$ whose output is always positive. It is defined as

$$f(x) = \begin{cases} x & \text{ii } x > 0 \\ 0 & \text{ii } x = 0 \\ -x & \text{ii } x < 0 \end{cases}$$

And graph is as follows.

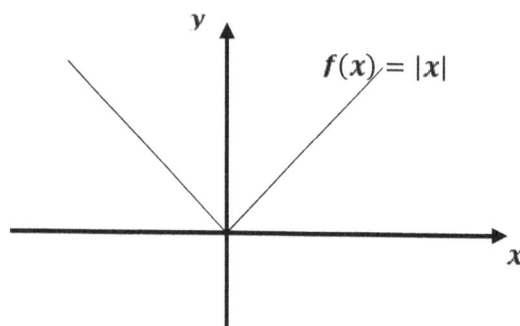

Solving the absolute function

To determine the roots of absolute function, we consider two cases, when the function is positive and when it is negative.

For instance, if the function is $f(x) = |x|$, and it is to be evaluated at $f(x) = c$, then we consider $-x = c$ and $x = c$.

Example 1

Find the value of x in the following functions
(i). $2|3 \quad -4| = 0$
(i). $1 + \frac{1}{3}|2 \quad +6| = 1$

<u>Solution</u>

(i). $2|3x - 4| = 0$

Diving through by 2, we get $|3x - 4| = 0$

By definition, we have $-(3x - 4) = 0$ and $3x - 4 = 0$

Thus, $-3x + 4 = 0$ and $3x - 4 = 0$
$-3x = -4$ and $3x = 4$; $x = \frac{4}{3}$ and $x = -\frac{4}{3}$

(ii). $1 + \frac{1}{3}|2x + 6| = 15$

Subtracting 1 on both sides, we have $\frac{1}{3}|2x + 6| = 14$; $|2x + 6| = 14 \times 3 = 42$

By definition of absolute value function, we have
$-(2x + 6) = 42$ and $2x + 6 = 42$
$-2x - 6 = 42$ and $2x + 6 = 42$
$-2x = 48$ and $2x = 36$
$x = -24$ and $x = 18$

RATIONAL FUNCTIONS

These are functions of the form $f(x) = \frac{h(x)}{g(x)}$ where both the numerator and denominator are polynomials and the denominator is not equal to zero.

Example, $f(x) = \frac{2x}{4x^2+1}$

The most important features of these functions are, the intercepts and the asymptotes.

Intercepts

These are points at which the function intercepts the $x-$ and $y-$ axes. At the $x-$intercept, $y = 0$, hence we have $\frac{h(x)}{g(x)} = 0$ implying $h(x) = 0$.

At $y-$ intercept, we have $x = 0$, hence we have $f(0) = \frac{h(0)}{g(0)}$ given that $g(0) \neq 0$.

Asymptotes

These are points at which a function gets close and closer to line but it does not touch it. There are three types, the vertical, horizontal and oblique asymptotes.

Vertical asymptote.

A function has a vertical asymptote a $x = t$ if $g(t) = 0$.

Horizontal and oblique asymptotes

The two does not occur at together.

If the degree of $h(x)$ is less than that of $g(x)$, then the horizontal asymptote is $x = 0$, but if they are equal, the horizontal asymptote is $\frac{a}{b}$ where a and b are the dealing coefficients of the functions, $h(x)$ and $g(x)$ respectively.

If the degree if $h(x)$ is more than that of $g(x)$ by 1, then the function has an oblique asymptote which is equal to the number of times, $g(x)$ divides $h(x)$ completely.

Example 2

Find the intercepts $f(x) = \frac{3\ +2}{6x^2-4}$

<u>Solution</u>

At the $x-$intercept, $y = 0$, hence $f(x) = \frac{3x+2}{6x^2-4} = 0$ impying that

$3x + 2 = 0$

Thus, $3x = -2$ and $x = -\frac{2}{3}$.

At the y-intercept, x=0, hence $f(0) = \frac{3(0)+2}{6(0)^2-4} = -\frac{2}{4} = -0.5$

Example 3

Find the asymptotes of $f(x) = \frac{x^2+2\ +2}{2\ -6}$

<u>Solution</u>

The vertical intercept occurs at x where $2x - 1 = 0, x = \frac{1}{2}$

The degree of the numerator if more than that of the denominator, hence, the function has oblique asymptote.

We carry out long division

$$
\begin{array}{r}
\frac{1}{2}x + \frac{5}{2} \\
2x-6 \overline{\smash{\big)}\ x^2 + 2x + 2} \\
\underline{-(x^2 - 3x)} \\
5x + 2 \\
\underline{-(5x - 15)} \\
17
\end{array}
$$

EXPONENTS

These are functions of the form $f(x) = a.b^k$ where a, b a k are constants with $b \neq 1, b \neq 0, a \neq 0$.

The graph of an exponential function is given as follows

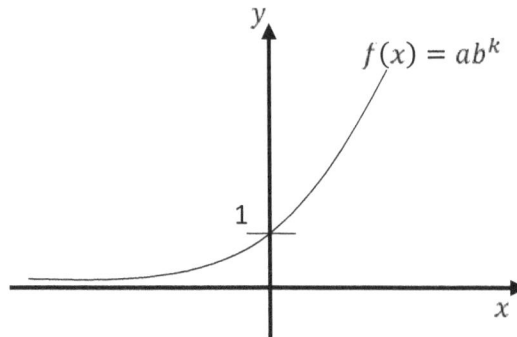

Given the data, one is required to fit an exponential model then use it to evaluate the function at given values of x or the independent variable.

Operation on exponents

The exponential functions can only be added when evaluated at a point. However, for multiplication and division, the use of laws of indices applied if the functions have a common base.

Logarithmic functions

These are functions of the form $f(x) = \log_a x$ where a is a nonzero constant that is not equal to one. These function are inverses of exponential functions.

The graph of a logarithm function is as shown below.

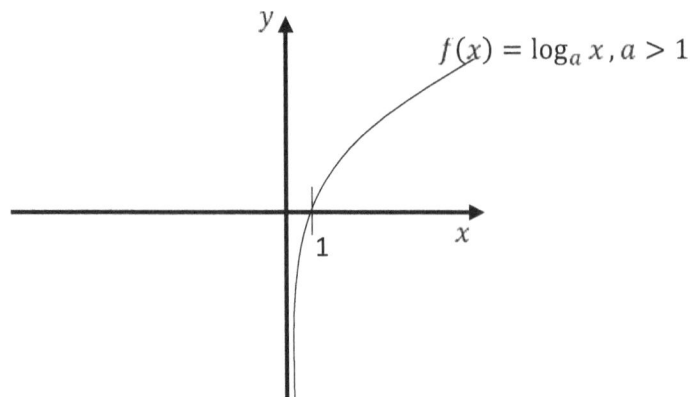

When the values of *a* is more than 1, the graph is as shown above. When it is between 0 and 1, the graph slants downwards as shown below.

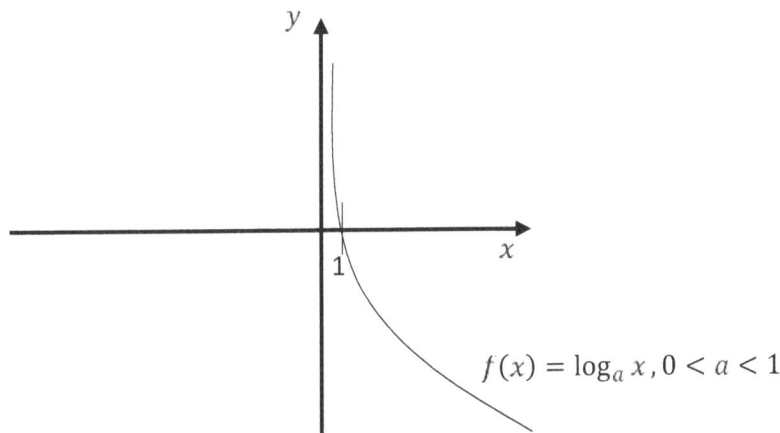

In practice, there are two most commonly used logarithm functions, the natural and the common logarithm. From the graph, it is clear that the function is one to one.

When the base, the value of *a=10*, we have the logarithm, $f(x) = \log_1 x$, being referred to as the common logarithm while when $a = e$, the Euler's numbers where $e = 2.718...$, the function is referred to as the natural logarithm and has a special notation, $f(x) = \log_e x = \ln x$.

Example 4

Given that $f(x) = \ln x, a$ what point is the value of the function a unit?

Solution

We find the value of x sich that $f(x) = 1$.

We have $f(x) = \ln x = 1$

Since, $\ln x = \log_e x$, we have $\log_e x = 1$.

In index notation, we have $x = e^1 = e$

Thus, when $f(x) = 1, x = e$.

ALGEBRA REVIEW, PARENT FUNCTIONS, ALGEBRAIC TRANSFORMATIONS, INVERSE FUNCTIONS

PARENT FUNCTION

A function is a very basic functional representation of any given function. For instance, the very basic quadratic function is $f(x) = x^2$. When other constants and linear functions comes it, they simply modify the function a process called algebraic transformation.

ALGEBRAIC TRANSFORMATIONS

This is a modification of the parent function. We may have translation, compression, stretch and reflection among others.

Translation (shift)

This refers to a slide of the parent function. When the parent functions $y = f(x)$ moves to the right by s, units, the resultant function is $y_1 = f(x - s)$ while when it moves by s units to the left, the resultant function is $y_2 = f(x + s)$.

When the function $y = f(x)$ shifts upwards by s units, the resultant is $y_1 = f(x) + s$ while when it shifts downwards by s units, the resultants is $y_2 = f(x) - s$.

Reflection

When a function $y = f(x)$ is reflected about the y −axis, the new function is $y_1 = f(-x)$ while when it is reflected about the x −axis, the resultant is $y_2 = -f(x)$

Stretch

When a function $y = f(x)$ is stretch vertically by a scale factor of r, we have $y_1 = r\ (x)$ where $r > 1$ while when stretched horizontally by a scale factor of r, we have $y_2 = f(r\ )$ where $0 < r < 1$.

Compression

When a function $y = f(x)$ is compressed vertically by a scale factor of r, we have $y_1 = r\ (x)$ where $0 < r < 1$ while when compressed horizontally by a scale factor of r, we have $y_2 = f(r\ )$ where $r > 1$.

INVERSE FUNCTION

A function, $f(x)$, is defined as a rule where one and only one element in the range is assigned an element in the domain. In inverse function we do the opposite. Its domain will be the range of the function, $f(x)$ while its range will be the domain of $f(x)$. The inverse of a function $f(x)$ is denoted $f^{-1}(x)$.

The function and its inverse are symmetrical at $y = x$.

To determine the inverse of $y = f(x)$, we make   the subject of the formula, then interchange the roles of y and x.

Example 1

Find the resultant function when $f(x) = 2x^2 + 3x - 1$ is shifted vertically upwards by 1 units, reflected about the y −axis, then compressed horizontally by scale factor of 3.

Solution

Vertically shift takes $y = f(x)$ to $y = f(x) + 1$, applying it, we get

$$y = f(x) + 1 = 2x^2 + 3x - 1 + 1 = f(x) = 2x^2 + 3x$$

Reflection about y −axis takes $y = f(x)$ to $y = f(-x)$, applying it, we get

$$y = f(-x) = 2(-x)^2 + 3(-x) = 2x^2 - 3x$$

Horizontal compression by a scale factor of 3 takes $y = f(x)$ to $y = f(3x)$, applying it, we get

$$y = f(3x) = 2(3x)^2 - 3(3x) = 18x^2 - 9x$$

The resultant function is $f(x) = 18x^2 - 9x$

Example 2

Given that function $y = 3(x + 2)^2 - 3(x + 2) - 3$ is a transformation of $f(x) = x^2 + x + c$, find the value of   and list all individual transformations involved.

<u>Solution</u>

First, we factorize out three since it is evident that $y = f(x)$ was multiplying by three. That gives as

$$y = 3(x + 2)^2 - 3(x + 2) - 3(1)$$

Also, it appears the function x was multiplied by a negative, so that we get $y = f(-x)$. Thus, we have

$$y = 3(-x - 2)^2 + 3(-x - 2) - 3(1)$$
$$y = 3((-x) - 2)^2 + 3((-x) - 2) + 3(-1)$$

We also find that the function was shifted 2 units to the right to get an equivalent of $y = f(x - 2)$

Thus, we have
$$y = 3((-x) - 2)^2 + 3((-x) - 2) + 3(1 - 2)$$
$$y = 3((-x) - 2)^2 + 3((-x) - 2) + 3(1 - 2)$$

Taking out the initial function, we get
$$y = x^2 + x + 1, c = 1$$

Hence, we have a vertical stretch by a scale factor of 3, relection about $y-$axis and a horizontal shift of 2 units to the right.

Example 3

Find the inverse of $f(x) = \frac{3x+1}{4x-7}$

<u>Solution</u>

$$f(x) = y = \frac{3x+1}{4x-7}$$

Making x the subject of the formula, we first cross multiply to get

$$4x - 7y = 3x + 1$$

$$4x - 3x = 7y + 1$$

$$x(4y - 3) = 7y + 1$$

$$x = \frac{7y+1}{4y-3}$$

Interchanging the roles of x and y, we have $f^{-1}(x) = \frac{7x+1}{4x-3}$

CONCLUSION

In this lesson, we began with vectors then moved to algebra where we defined a function, looked at its properties. We then moved to looking at types of functions and their transformations. We have finalized by looking at the inverses of functions which are functions that maps a range of a function back to the domain.

CONICS, CORDINATE SYSTEM, SERIES AND SEQUENCES

WHAT YOU NEED TO KNOW

In this session, we will discuss the properties and features of parabolas, hyperbolas, ellipse and circles generally known as ellipse. We will also look at polar and parametric equations as one of the alternative ways of describing equations of lines and surfaces. We will finalize by looking at series, sequences in relation to limits as well as limits of other functions.

MATH TOPICS

- Pre-calculus 103.11 Conics (Reference 1.11).
- Pre-calculus 103.12 Polar and Parametric (Reference 1.12).
- Pre-calculus 103.13 Sequences, series and limits. (Reference 1.13).

INTRODUCTION

We usually meet activities or events that happen after a certain period of time such that the history of their occurrence becomes sequences and series. An important thing to note here is that their occurrence can be predicted at a certain time using the concepts of limits. We also use mirrors, parabolic, we meets circles and ellipse every day and it is of importance if we understand some of the basic features of these so that we can utilize the optimally. Our lesson will be based on the conics, the polar and parametric equations as well as the series, sequences and limits.

CONICS

These are curves generated when one or two cones are cut at different angles. They are circles, parabola, ellipse and hyperbola.

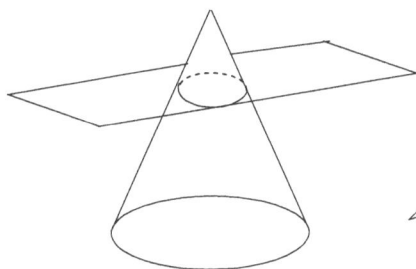
Circle

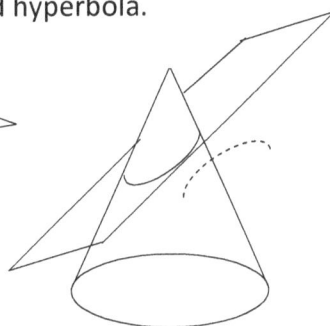
Ellipse

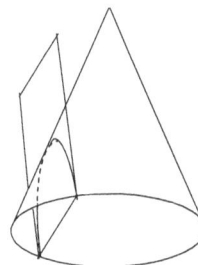
Parabola

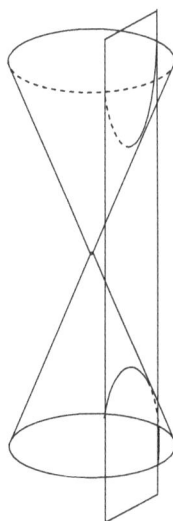
Hyperbola

CIRCLE

It is the set of points that are at an equal distance from a fixed point called the center of the circle. The standard equation of the circle is given by

$$Ax^2 + By^2 + C + D + E = 0$$

This equation can be reduced by completing square to center-radius form as $(x - a)^2 + (y - b)^2 = r^2$ where (a, b) is the center and r the radius.

PARABOLA

It is the set of points that are far from the central point, called the focus, as the line called the directrix.

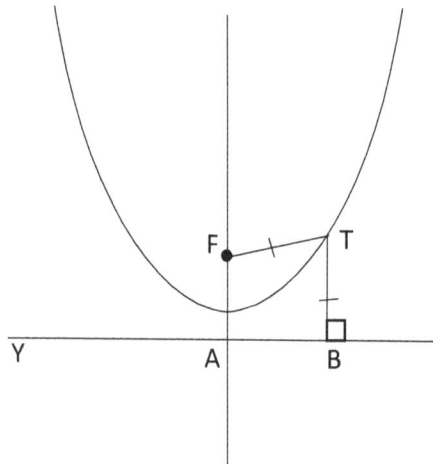

The curve above is a parabola. F is the focus and YAB the directrix. T represents any point on the parabola. T is such that $F = T$. The line AF divides the parabola into two equal and overlapping parts hence, it is the line of symmetry.

The equation of the parabola facing upwards is given by

$$(x - a)^2 = 4p(y - b)$$

Where, (a, b) is the vertex, p is the distance from the vertex to the focus and $y = b - p$ the directrix.

When the parabola is facing downwards, we have

$(x - a)^2 = -4p(y - b)$ and the directrix as $y = b + p$

When it is facing to the right, the equation is $(y - a)^2 = 4p(x - b)$ where (a, b) is the vertex and   the distance from the vertex to the focus and $x = a - p$ the directrix.

When facing to the left, the equation is is $(y - a)^2 = -4p(x - b)$ where $x = a + p$ is the directrix.

ELLIPSE

An ellipse is a collection of points such that sum of the distances from each point to two fixed points called the focus is the same for all points. The equation of the ellipse is given by

$$\frac{(x - r)^2}{a^2} + \frac{(y - t)^2}{b^2} = 1$$

Where (r, t) is the center of the ellipse, a and b are the distances from the center to the $x-$ and $y-$ intercept respectively.

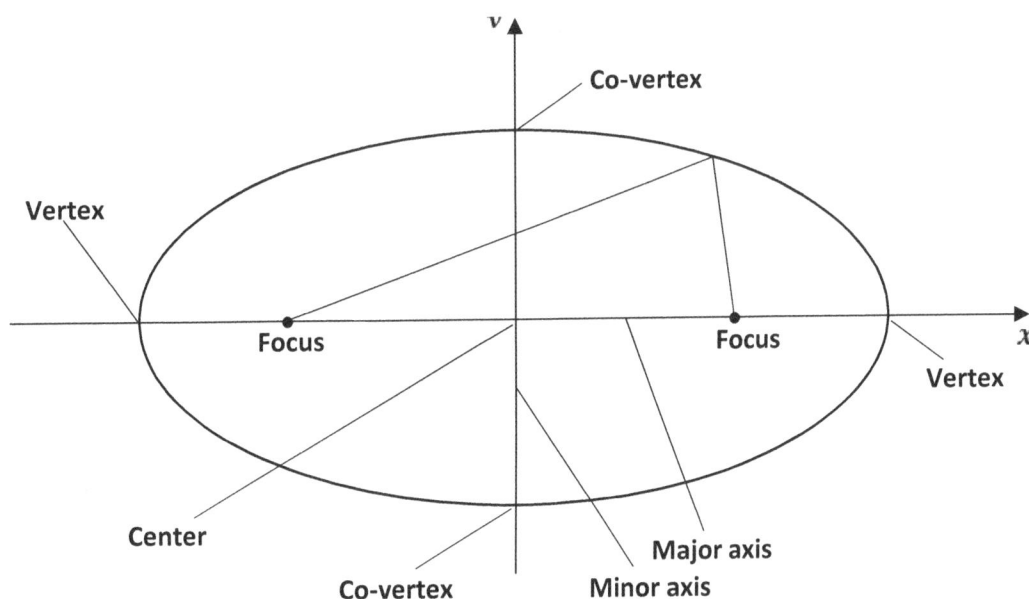

If the major axis is on the y−axis, then the equation becomes

$$\frac{(y-r)^2}{a^2} + \frac{(x-t)^2}{b^2} = 1$$

Where (r, t) is the center of the ellipse, a and b are the distances from the center to the y − and x − intercept respectively.

The distance from the center to the foci is c, where $c^2 = b^2 - a^2$

HYPERBOLA

It is a collection of points such that the positive difference between the distances from the one point to the fixed points called the foci is the same. Each of the two curves that make up a hyperbola is called a Branch. Each Branch has a vertex that lie on one line together with the foci of the hyperbola. The line where the foci and the vertices lie is called the transverse axis.

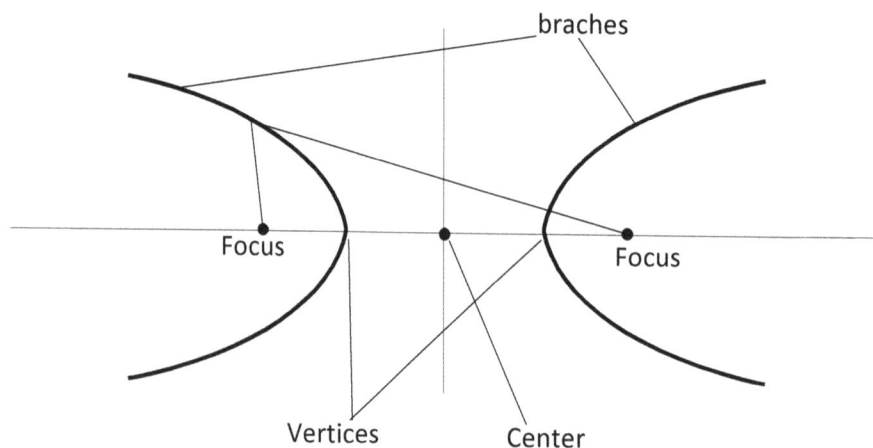

The equation of a hyperbola is given by

$\frac{x^2}{a^2} - \frac{y^2}{b^2} = 1$ When it is horizontal. In that case, its center is at the origin and the vertices are   units from the center.

The asymptotes of the hyperbola are $y = \pm\frac{a}{b}$, whether it is horizontal or vertical.

When the hyperbola is vertical, its equation is $\frac{y^2}{b^2} - \frac{x^2}{a^2} = 1$ where its center is at the origin and the vertices are b units from the center.

When the hyperbola has the center at (s, t), where then its equations are $\frac{(x-s)^2}{a^2} - \frac{(y-t)^2}{b^2} = 1$ and $\frac{(y-s)^2}{b^2} - \frac{(x-t)^2}{a^2} = 1$ for horizontal and vertical hyperbolas respectively.

The distance from the center to the foci is c, where $c^2 = b^2 - a^2$.

Example 1
Find the equation of a parabola whose vertex is (2,1) and directrix is $x = -3$.

Solution

The directrix is parallel to the y–axis showing that the parabola is horizontal. The vertex is on the right of the directrix showing that the parabola is facing the positive side of the x–axis hence, p is positive.

Thus, the equation is given by
$$(y - a)^2 = 4p(x - b)$$

Where, (a, b) is the vertex, p is the distance from the vertex to the focus. $a = 2, b = 1$, thus, we need to know p.

Since the symmetric line is horizontal, p, is the horizontal distance. This makes us consider x–coordinates only. From $(2,1)$ to $x = -3$, it is $|1 - -3| = 4\, u\, i$. Thus, $p = \frac{4}{2} = 2$

The equation is $(y - 1)^2 = 8(x - 2)$

Example 2

What would be the directrix of a parabola given by $(y-1)^2 = -2$

Solution

Comparing with the general equation of a parabola whose vertex is (s,t), we get that its center is at $(1,0)$.

Comparing the equation with $y^2 = 4p$, we get that $4p = -2x, p = -\frac{1}{2}$

Since the y is squared in the equation, the parabola is horizontal. Since p is negative, the parabola is facing in the negative side of $x-$axis. Thus, the directrix is 0.5 units to the right of the vertex.

Thus, the directrix is $x = 1 + 0.5$; $x = 1.5$

Example 3

Find the foci of the ellipse $x^2 + 4y^2 = 4$.

Solution

If we divide through by 4, we get $\frac{x^2}{4} + y^2 = 1$

Thus, we have $\frac{x^2}{2^2} + \frac{y^2}{1^2} = 1$, implying that, $a = 2$ and $b = 1$

Since the equations is of the form, $\frac{x^2}{a^2} + \frac{y^2}{b^2} = 1$, its center is at the origin.

Since $a > b$, the ellipse is horizontal. The major axis is parallel to the

$x - a$.

The foci are c units from the center, where $c^2 = a^2 - b^2 = 2^2 - 1 = 3$;

$c = \sqrt{3}$

The foci are $\sqrt{3}$ units from the center, hence, they are at $(\sqrt{3}, 0)$ and $(\sqrt{3}, 0)$

Example 4

Find the radius and the center of the circle $3x^2 + 6 \ + 3y^2 - 3 = 1$

<u>Solution</u>

We divide through by 3 the complete the square,

$3x^2 + 6x + 3y^2 - 3 = 12y$ is equivalent to $x^2 + 2x + y^2 - 4y = 1$

$x^2 + 2x + 1 - 1 + y^2 - 4y + 2^2 - 2^2 = 1$

$x^2 + 2x + 1 + y^2 - 4y + 2^2 = 1 + 1 + 4$

$(x - 1)^2 + (y - 2)^2 = 6$

$(x - 1)^2 + (y - 2)^2 = \left(\sqrt{6}\right)^2$

Thus, the center is $(1,2)$ and the radius is $\sqrt{6}$

Example 5

Find the focus and the hyperbola given by $5(y - 1)^2 - 9x^2 = 4$.

<u>Solution</u>

We convert the equation into the standard form. Dividing through by 45, we get

$$\frac{(y - 1)^2}{9} - \frac{x^2}{5} = 1$$

Since the term with y is positive, the hyperbola is vertical, hence $b^2 = 9, a^2 = 5$

The focus c units from the center. The center is at $(0,1)$

We have $c^2 = b^2 - a^2; c = \sqrt{b^2 - a^2} = \sqrt{9 - 5} = \sqrt{4} = 2$

Since the hyperbola is vertical, the foci are at $(0,1 + 2) = (0,3)$ and $(0,1 - 2) = (0,-1)$.

POLAR AND PARAMETRIC

We are used to equations where a functions expressed in terms of one independent variable, mostly, x. However, there are cases where it is necessary to express both the independent and the dependent variables in terms on yet another variable. Such equations are called parametric equations.

For instance, if we want to find the distance moved by a particle in space, since from one point, the particle moves making a displacement along the horizontal axis (x) as well as the vertical axis, (y), we may need to express these displacements as a function of times, thus, we have the distances,

$y = f(t)$ Vertical displacement

$x = g(t)$Horizontal displacement, t, is the parameter, in this case. The parameter is defined by the bounds, that is, the lower and the high limits that it can get to.

Example 1
1. the motion of a particle is given by $y = x^3 + 4x^2 - 3 \ + 5$. Express y and x interms of t if $x = 2$. Where $0 < t < 0.5$.

<u>Solution</u>

Since, $x = 2t$, we already have the function, $x(t) = 2t$

Since, $y = x^3 + 4x^2 - 3x + 5$, we substitute x with $2t$ to get $y = (2t)^3 + 4(2t)^2 - 3(2t) + 5$
$y = 8t^3 + 8t^2 - 6t + 5$

The parametric equations are
$x(t) = 2t$
$y(t) = 8t^3 + 8t^2 - 6t + 5 , 0 < t < 0.5$

The most common parametric equations are the equations involving the angle whose variable θ (s᷍).

We consider a ray whose tail is at the origin and the head is at (x, y). Let the ray make an angle θ with the positive $x-$axis. We would like to find out how we can express x and y interm of the angle, θ.

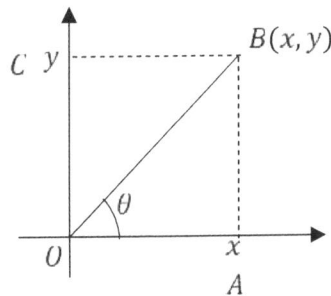

Using trigonometric ratios, we have

$$\cos\theta = \frac{O}{O}; O = O \cos\theta \quad ....(i)$$

$$\sin\theta = \frac{A}{O}; \quad O = A = O \sin\theta..............(ii)$$

Using distance from the origin, $O = x, O = y, O = \sqrt{x^2 + y^2}$

Using the equations of a circle of radius 1, we get that
$$x^2 + y^2 = r^2 = 1, he \quad \sqrt{x^2 + y^2} = r$$

Upon substitution, we get that
$$x = r\cos\theta, y = r\sin\theta$$

But it is only the angle that changes but radius, r, remains constant. Since the angle can change from $0°$ to $360° = 2\pi$ radians, we have the parametric equations

$$x = r\cos\theta$$
$$y = r\sin\theta, \quad 0 < \theta < 2\pi$$

Note that

$$\frac{y}{x} = \frac{r\sin\theta}{r\cos\theta} = \frac{\sin\theta}{\cos\theta} = \tan\theta; \text{thus } \theta = \tan^{-1}\left(\frac{y}{x}\right)$$

POLAR EQUATIONS

Polar equations are equations where the functions, radius, r, are expressed in terms of the angle. The coordinate system where the location of a point is in terms of x and y is called the rectangular coordinates while that where the location is in terms of angle and the radius (length of a ray) is called the **polar coordinates.**

Thus, from rectangular coordinates to polar coordinates, we have

$$x = r\cos\theta$$

$$y = r\sin\theta\ , \qquad 0 < \theta < 2\pi$$

$$r = \sqrt{x^2 + y^2}$$

$$\text{and } \theta = \tan^{-1}\left(\frac{y}{x}\right)$$

The polar coordinates are given as (r, θ) where $r = \sqrt{x^2 + y^2}$ and $\theta = \tan^{-1}\left(\frac{y}{x}\right)$.

Example 1
Write $(3, 4)$ and $(2, 2)$ in polar coordinates.

Solution

(3,4)

$$x = 3, y = 4, r = \sqrt{x^2 + y^2} = \sqrt{3^2 + 4^2} = 5$$

$$\theta = \tan^{-1}\left(\frac{y}{x}\right) = \tan^{-1}\frac{4}{3} = 53.13°$$

The coordinates are $(5, 53.13°)$

(2,2)

$$x = 2, y = 2, r = \sqrt{x^2 + y^2} = \sqrt{2^2 + 2^2} = \sqrt{8} = 2\sqrt{2}$$

$$\theta = \tan^{-1}\left(\frac{y}{x}\right) = \tan^{-1} 1 = 45°$$

The coordinates are $\left(2\sqrt{2}, 45\right)$ o $\left(2\sqrt{2}, \frac{\pi}{4}\right)$

Example 2

Convert $(2, \frac{2}{3}\pi)$ to rectangular coordinates

Solution

We have $r = 2, \theta = \frac{2}{3}\pi = 120°$

$$x = r\,c \quad \theta = 2\cos 120° = -1$$
$$y = r\,s \quad \theta = 2\sin 120° = 1.732$$

The coordinates are $(-1, 1.732)$

Example 3.

Change the following equations to polar equation.

(i). $x^2 + y^2 = 2$

(ii). $\frac{x^2}{4} + \frac{y^2}{1} = 1$

(ii). $x = 2 + x^2$

Solution

(i). Since $r = \sqrt{x^2 + y^2}$ or $x^2 + y^2 = r^2$, we have $r^2 = 25$

Thus, the equation of c circle of radius 5 is $r = 5$.

(ii). We have $= r\cos\theta$, $y = r\sin\theta$. Squaring each case, we have $x^2 = r^2\cos^2\theta$ and $y^2 = r^2\sin^2\theta$

Substituting into the above equation, we have

$$\frac{r^2\cos^2\theta}{4} + \frac{r^2\sin^2\theta}{16} = 1$$

Multiply through by 16, we have $4r^2\cos^2\theta + r^2\sin^2\theta = 16$

Or

$$r^2(4\cos^2\theta + \sin^2\theta) = 16$$
$$r^2 = \frac{16}{4\cos^2\theta + \sin^2\theta}$$
$$r = \sqrt{\frac{16}{4\cos^2\theta + \sin^2\theta}} = \frac{4}{\sqrt{4\cos^2\theta + \sin^2\theta}}$$

$$r = \frac{4}{\sqrt{4\cos^2\theta + \sin^2\theta}}$$

But, we know that $\cos^2\theta = \frac{1}{2} + \frac{1}{2}\cos 2\theta$; $\sin^2\theta = \frac{1}{2} - \frac{1}{2}\cos 2\theta$

Upon substitution, we get

$$r = \frac{4}{\sqrt{4\cos^2\theta + \sin^2\theta}} = \frac{4}{\sqrt{4\left(\frac{1}{2} + \frac{1}{2}\cos 2\theta\right) + \frac{1}{2} - \frac{1}{2}\cos 2\theta}}$$

$$r = \frac{4}{\sqrt{2 + 2\cos 2\theta + \frac{1}{2} - \frac{1}{2}\cos 2\theta}} = \frac{4}{\sqrt{2.5 + 1.5\cos 2\theta}}$$

Thus, the equation is $r = \dfrac{4}{\sqrt{2.5 + 1.5\,c\quad 2\theta}}$

(ii). $x = 2 + x^2$

$x = r\cos\theta$, $y = r\sin\theta$, $x^2 = r^2\cos^2\theta$, $x = r^2\cos\theta\sin\theta$

Substituting into the above equation, we get

$$r^2\cos\theta\sin\theta = 2 + r^2\cos^2\theta$$
$$r^2\cos\theta\sin\theta - r^2\cos^2\theta = 2$$
$$r^2\cos\theta(\sin\theta - \cos\theta) = 2$$

Example 4

Convert the following to rectangular equations

(i). $r = 2\quad \theta + 1$

(ii). $r = 4\quad \theta$

<u>Solution</u>

We have $r = \sqrt{x^2 + y^2}$, $x = r\cos\theta$, $y = r\sin\theta$, $\cos\theta = \frac{x}{r}$, $\sin\theta = \frac{y}{r}$

(i).Thus, $r = 2c\quad \theta + 1$, is equivalent to

$$r = 2\frac{x}{r} + 1$$

Multiplying through by r, we get $r^2 = 2x + r$

Upon substitution for r, get

$$x^2 + y^2 = 2x + \sqrt{x^2 + y^2}$$
$$o$$
$$x^2 + y^2 - 2x = \sqrt{x^2 + y^2}$$

Squaring both sides, we get $(x^2 + y^2 - 2x)^2 = x^2 + y^2$

Simplifying further, we get

$$x^4 + y^4 + 2x^2y^2 - 4x^3 - 4xy^2 + 4x^2 = x^2 + y^2$$
$$x^4 + y^4 + 2x^2y^2 - 4x^3 - 4xy^2 + 3x^2 - y^2 = 0$$

Thus, $r = 4\sin\theta$, is equivalent to $r = \frac{4y}{r}$
$$r^2 = 4y$$

Upon substitution, we get $x^2 + y^2 = 4y$

SEQUENCES, SERIES AND LIMITS

A sequence is a pattern of numbers generated using a given formula or criteria. The numbers in the sequence are called terms. The representation where the terms are added is called a series.

A sequence is denoted by $\{a_n\}_{n-1}^{\infty}$ sometimes as $a_n, n \geq 1$. The most common sequences are geometric and arithmetic sequences. The general term is a sequence is referred to as the nth term.

An arithmetic sequence is a sequence where the difference between any two consecutive terms is the same. This difference is called the common difference abbreviated as d. If a, is the first term of an arithmetic sequence, then the nth term, $a_n = a + (n - 1)d$

Geometric sequence is a sequence where the quotient (ratio) between a term and its preceding one is always the same. This ratio is called the common ratio and it is abbreviated as r. If the first term is a and the common ratio is r, then the nth term is given by $a_n = ar^{n-1}$.

SERIES

It is the sum of terms in a given sequence. In most cases, we use a notation to denote such summation. When we want to sum a given number of terms of a sequence, we use the notation called sigma notation, Σ.

If we have the numbers $a_1, a_2, a_3, a_i \dots a_{n-1}, a_p$ and we wish to sum from a_i to a_n, that is $a_i, a_{i+1}, a_{i+2}, a_{i+3} \dots a_{n-2}, a_{n-1}, a_p$ where i is a number less than n, we have

$$a_i + a_{i+1} + a_{i+2} + a_{i+3} + \cdots + a_{n-2} + a_{n-1} + a_p = \sum_{n-i}^{p} a_n$$

In this notation, the smallest number where we begin the notation is called the lower limit of summation, i, while the highest number, p, is called the upper limit of summation.

For instance, if we have $\sum_{n-1}^{3} 3n$, we have

$$\sum_{n-1}^{3} 3n = 3(1) + 3(2) + 3(3) = 3 + 6 + 9 = 18$$

Operations with summation notation

When we have two series, $\sum_{n-1}^{p} a_n$ and $\sum_{n-1}^{p} b_n$ then we have

$$(i). \quad \sum_{n-1}^{p} a_n \pm \sum_{n-1}^{p} b_n = \sum_{n-1}^{p} (a_n \pm b_n)$$

$$(ii). \quad k \sum_{n-1}^{p} b_n = \sum_{n-1}^{p} k_n$$

(ii). if m is between i and p, then $\sum_{n-i}^{m} b_n + \sum_{n-m}^{p} b_n = \sum_{n-i}^{p} b_n$

$$(ii). \quad \sum_{n-i}^{p} b_n = \sum_{n-i+1}^{p+1} b_{n-1}$$

Arithmetic series

Arithmetic series is the sum of the first n terms of an arithmetic sequence. The sum of first n terms of an arithmetic series is given by

$$s_n = \frac{n}{2}(a + a_n) \text{ where } a_n = a + (n-1)d$$

Geometric series is the sum of the first terms of first, n, terms of a geometric sequence. This sum is given by $s_n = \frac{a(1-r^n)}{1-r}$.

Limits in series

A limit of a function, is a number the function gets close and close to as the inputs of the function is chosen close to a given number. The limit, of a sequence is the number the sequence gets to as the value of n is close and close to a given number.

For instance, the limit of the sequence s_n, as n approached 100 is the number that $s_n = \frac{a(1-r^n)}{1-r}$ gets closer and closer to when we substitute for $n = 97, 98, 99, 99.5$, e .

In series, the most important application of limit is limit to infinity of a geometric sequence.

If $r < 1$, then the limit as n approaches infinity is the value that $s_n = \frac{a(1-r^n)}{1-r}$ gets to as n becomes bigger and bigger. Since $r < 1$, r^n becomes smaller and smaller and smaller as it tends to zero.

Thus the limit denotes, $\lim_{n\to\infty} r^n = 0$

Therefore

$$\lim_{n\to\infty} s_n = \lim_{n\to\infty} \frac{a(1-r^n)}{1-r} = \lim_{n\to\infty} \frac{a(1-0)}{1-r} = \frac{a}{1-r}$$

Example 1
Compute the sums

    (i). $\sum_{n-1}^{5} 2n^2$

    (ii). $\sum_{n-1}^{4} \frac{3}{n}$

Solution
We substitute for each value of n

    (i). $\sum_{n-1}^{5} 2n^2 = 2(1)^2 + 2(2)^2 + 2(3)^2 + 2(4)^2 + 2(5)^2 = 2 + 8 + 18 + 32 + 50 = 110$

    (ii). $\sum_{n-1}^{4} \frac{3}{n} = \frac{3}{1} + \frac{3}{2} + \frac{3}{3} + \frac{3}{4} = 36 + 18 + 12 + 9 = 75$

Example 2

Find the expression for the nth term of the following sequence, 5, 7, 9, 11,…

<u>Solution</u>

This is a geometric series because the difference between any two terms is 2.

Hence $d = 2$

The first term, $a = 5$

The nth term is $a_n = a + (n-1)d = 5 + 2(n-1) = 5 + 2n - 2 = 2n + 3$

Example 3

Compute the 4th and 10th terms of an arithmetic sequence whose common difference is 3 and first term is -17.

<u>Solution</u>

The first term is $a = -17$ and the common difference $d = 3$

Using the formula, $a_n = a + (n-1)d$, we have

$$a_4 = -17 + 3(4-1) = -8$$
$$a_1 = -17 + 3(10-1) = 10$$

Example 4

Compute the 3th and 5th terms of a geometric sequence whose common ratio is -0.5 and first term is 24.

<u>Solution</u>

The first term is $a = 24$ and the common ratio is $r = -0.5$
The nth term is given by $a_n = ar^{n-1}$
Thus, the 3rd and 5th terms are
$$a_3 = 24 \times (-0.5)^{3-1} = 24 \times (-0.5)^2 = 24 \times 0.25 = 6$$
$$a_5 = 24 \times (-0.5)^{5-1} = 24 \times (-0.5)^4 = 24 \times 0.0625 = 1.5$$

Example 5

A ball bounces back on a flat ground after a distance of 4 feet and 3.2 feet in the first and second bounce respectively. Determine the total distance covered by the ball along the ground if it continued bouncing back and forth at the same rate without any disturbance.

<u>Solution</u>

The first bounce is 4 ft and the second one is 3.2 ft

Thus, we have the first term $a = 4$ and the second term $a_2 = 3.2$

Since the ball will continue at the same rate, the distances between successive bounces have a common quotient, $r = \frac{a_2}{a} = \frac{3.2}{4} = 0.8$

Since it bounces to infinite number of times, we use the summation

$$\lim_{n \to \infty} s_n = \lim_{n \to \infty} \frac{a(1 - r^n)}{1 - r} = \frac{a}{1 - r} = \frac{4}{1 - 0.8} = 20 \, f$$

CONCLUSION

In this lesson, we have discussed the conics by looking at different ways they are generated. We also looked at their features and their standard formula. We then proceeded to discussing parametric equations with special attention to polar equations. We have completed by looking at the introduction to sequences and series of numbers. Some important concepts discussed include the sigma notation and the geometric and arithmetic series that are highly applicable in real life.

GLOSSARY

Absolute function: This is a function what has all or some of its terms enclosed in absolute symbols, | |

Amplitude: The distance from central line to highest point of the periodic graph.

Angular frequency: The number of times the waves makes complete circles in every 2π.

Arithmetic Sequence: It is a sequence where the difference between any two consecutive terms is the same.

Arithmetic Series: Arithmetic series is the sum of the first n terms of an arithmetic sequence.

Asymptotes: These are points at which a function gets close and closer to line but it does not touch it.

Bijective function: Function that is both surjective and injective.

Binomial: polynomial is composed of two terms.

Circle: It is the set of points that are at an equal distance from a fixed point called the center of the circle.

Columnvector: A vector whose tails is not at the origin.

Conics: These are curves generated when one or two cones are cut at different angles.

Cubic function: A polynomial with degree 3.

Decreasing function: A function, $f(x)$ is decreasing if for any two values, c and d such that $c < d$, then the output with respect to c must be more than the output with respect to d.

Degree-Minutes-Seconds system (DMS): A system of measuring angles that uses degrees.

Degree of a polynomial: The number n which is the highest power in the expression.

Domain: The set of all inputs of a function.

Ellipse: An ellipse is a collection of points such that sum of the distances from each point to two fixed points called the focus is the same for all points.

Exponents: These are functions of the form $f(x) = a.b^{kx}$ where a, b a k are constants with $b \neq 1, b \neq 0, a \neq 0$.

Even function: A function $f(x)$ is even if substituting x with $-x$ gives $f(x)$.

Factors of a polynomial: It is an expression that divides the polynomial completely.

Function: It is a relation where every element of the input corresponds to only one members of the output.

Geometric Sequence: It is a sequence where the quotient (ratio) between a term and its preceding one is always the same.

Geometric Series: It is the sum of the first terms of first, n, terms of a geometric sequence.

Hyperbola: It is a collection of points such that the positive difference between the distances from the one point to the fixed points called the foci is the same.

Increasing function: A function, $f(x)$ is increasing when for any two values in the domain, a and b such that b is more than a, then the output with respect to b, $f(b)$ must be more than that of a, $f(a)$.

Injective function: Functions that have a one to one relation between the elements of the domain and that of the range.

Interval Phase shift: The number of units the graph moves horizontally from the parent graph.

Intercepts: These are points at which the function intercepts the $x-$ and $y-$ axes.

Inverse function: A function that maps a range of a function to the domain.

Limit of a Function: A limit of a function, is a number the function gets close and close to as the inputs of the function is chosen close to a given number.

Line of symmetry: the line that divides a curve into two equal and overlapping parts.

Linear function: A polynomial with degree 1.

Local minima: A function $f(x)$ has a local minima at a point say $x = a$, if $f(x)$ is decreasing for points close and on the left of $x = a$ and increasing for points close and on the right of $x = a$.

Logarithmic functions: These are functions of the form $f(x) = \log_a x$ where a is a nonzero constant that is not equal to one.

Modulus of a vector: It is the length of a vector.

Monomial: polynomial is composed of one term.

Multi-angle trigonometry: Is a concept where a trigonometric problem has a solution that are very many angles.

Odd function: A function $f(x)$ is odd if substituting x with $-x$ gives $f(x)$.

Parallelvectors: Vector are parallel, when their numeric representation is a scalar multiple of the other.

Parabola: It is the set of points that are far from the central point, called the focus, as the line called the directrix.

Parent function: A function is a very basic functional representation of any given function.

Parametric Equations: Equations where variables are expressed in terms of another independent new variables.

Period: number of times a cycle is completed.

Polar Equations : Polar equations are equations where the functions, radius, r, are expressed in terms of the angle.

Polynomial: It is a function that is generally given by $p(x) = a_0 + a_1 x + a_2 x^2 + \cdots + a_n x^n$ where, $a_0, a_1, \ldots a_n$ are called coefficients, $a_n \neq 0$, is called the leading coefficient and x is the variable.

Position vector: Vector whose tail is at the origin.

Quadratic function: A polynomial with degree 2.

Quartic function: A polynomial with degree 4.

Radian: Is equal to a central angle whose arc length is equal to the radius.

Range: The set of all outputs of a function.

Rational functions: These are functions of the form $f(x) = \frac{h(x)}{g(x)}$ where both the numerator and denominator are polynomials and the denominator is not equal to zero.

Rectangular Coordinates: The coordinate system where the location of a point is in terms of x and y is Polar coordinates: This is where the location is in terms of angle and the radius (length of a ray).

Roots of a polynomial: These are values of a variable which when substituted into the polynomial, it reduces it to zero.

Sequence: It is a pattern of numbers generated using a given formula or criteria.

Series : It is the sum of terms in a given sequence.

Surjective function: Functions where for every output, there is a corresponding input.

Trigonometric identities: They are a number of trigonometric rules set to help solve trigonometric expressions and equations.

Trigonometric equations: These are algebraic representations where trigonometric expressions are equated.

Trigonometric ratios: Are ratios of two sides of a triangle with relation to a given reference angle.

Unit circle: Is a circle of radius 1. As used in this lesson, it has to be centered at the origin.

Vector: It is a geometric entity that has direction as well as quantity.

Vertical shift: The distance moved vertically with respect to the parent graph.

www.ingramcontent.com/pod-product-compliance
Lightning Source LLC
Chambersburg PA
CBHW051420200326
41520CB00023B/7311